Elementary Algebra

A Gentle Approach

By

Leonard Sperduto

This book is dedicated to James Lawson

CONTENTS

4

CHAPTER 1
ENTERING THE WORLD OF ALGEBRA

SECTION 1.1
TRANSLATIONS

Students must learn to translate the written language into mathematical symbols upon entering the realm of Algebra. The basic operations of addition, subtraction, multiplication, and division all have written words that translate into the appropriate symbol. Sum, plus, increased by, and more than are words that mean addition. Students should be familiar with the idea that 6 plus 9 is the same as $6 + 9$. In Algebra, numbers can be represented by any letter of the alphabet. H plus W means $H + W$.

EXAMPLE 1:

a)   The sum of L and 7 translates to $L + 7$.

b)   3 plus h translates to $3 + h$.

c) K increased by Y translates to $K + Y$.

d) g more than b translates to $b + g$. ■

Notice that Example 1d translates in a different order than the other examples. In this example, b is the starting point, and g is being added to b. A number is usually added to a starting point. The translations for addition follow the pattern starting point plus whatever is being added.

Difference, minus, decreased by, and less than are words that mean subtraction.

EXAMPLE 2:

a)   The difference of n and d translates to $n - d$.

b) V minus 1 translates to $V - 1$.

c)   2 decreased by w translates to $2 - w$.

d) J less than B translates to $B - J$. ■

Notice that example 2d translates in the same manner as example 1d. The reasoning is the same. These translations follow the same pattern as addition. The starting point minus whatever is being subtracted.

Product and times are two words that mean multiplication. In Algebra, the conventional symbol of $\times$ can no longer be used to represent multiplication. Any letter of the alphabet can be used to represent any number, and the symbol $\times$ looks like the letter x; therefore, different symbols must be used to represent multiplication. A raised dot •, parentheses ( ), and numbers and or letters placed next to each other all indicate multiplication.

EXAMPLE 3:

a)   4 times 3 translates to 4 • 3

b)   *P* times *M* translates to *PM*.

c)   The product of 2 and 10 translates into (2)(10).

d)   5 times *a* translates to 5*a*.

e)   The product of 6, *n*, and *f* translates to 6*nf*. ■

Quotient and divided by are two terms that mean division. The conventional symbol of ÷ is still valid in Algebra. A fraction can also be used to represent division.

EXAMPLE 4:

a)   *Q* divided by 8 translates to *Q* ÷ 8.

b)   The quotient of *L* and *S* translates to $\dfrac{L}{S}$. ■

The first four examples showed how each of the four basic operations translates from words to symbols. The next example will show how combinations of the operations translate.

EXAMPLE 5:

a)   The sum of *b* and 1 times the sum of *e* and 2 translates to $(b + 1)(e + 2)$.

b)   4 times the difference of *c* and *d* translates to $4(c - d)$.

c)   The sum of *u* and *k*, divided by 6 translates to $\dfrac{u + k}{6}$.

d)   The difference of *y* and *z*, divided by the sum of *y* and *z* translates to $\dfrac{y - z}{y + z}$. ■

There is one more operation that needs to be mentioned before going further into the world of Algebra. The exponential operation also has words that translate to symbols.

Square means to raise to the second power, and cube means to raise to the third power. Exponentials can be combined with other operations. The meaning of exponents and how exponents are used is explained in the next section.

EXAMPLE 6:

a)   The square of w translates to w^2.

b)   The cube of m translates to m^3.

c)   The sum of j squared and o translates to $j^2 + o$.

d)   The difference of k cubed and p translates to $k^3 - p$.

e)   The product of n squared and q translates to $n^2 q$.

f)   The quotient of r and s cubed translates to $\dfrac{r}{s^3}$. ∎

PROBLEMS:

Translate the following statements into algebraic symbols.

1)   The sum of d and f.

2)   18 more than p.

3) j minus k.

4) a decreased by 2.

5)   3 less than d.

6) f times h.

7)   The product of 6 and g.

8)   The product of 8, j, and k.

9)   2 times the sum of c and d.

10)   The sum of b and c times the difference of b and c.

11)   The product of c and 4 less than c.

12) a divided by 2.

13)   The quotient of c plus d, divided by 2.

14)   The difference of d and f, divided by 5.

15)   The sum of b and 5, divided by the difference of b and 5.

16)   The sum of f squared and k, divided by 29.

17)   The difference of b squared and c.

18)   The sum of a cubed and b.

SECTION 1.2
EXPONENTS AND ORDER OF OPERATIONS

Before learning order of operations, one must learn about exponents first. Exponents are a part of order of operations. An exponent, in its simplest form, is a number that is located above and to the right of a number or unknown as shown in Example 6 in Section 1.1. <u>An exponent tells us how many times a number or unknown, known as the base, multiplies itself.</u> A base with an exponent is known as an exponential form.

EXAMPLE 1:

a) $5^3 = 5 \bullet 5 \bullet 5 = 125$ 3 is the exponent, 5 is the base.

b) $D^4 = DDDD$ 4 is the exponent, D is the base. ■

WARNING: A common error is to multiply the base by the exponent. $5^3 \neq 5 \bullet 3$. Example 1a shows the result of 5^3 is 125. $5 \bullet 3 = 15$.

In the previous example, we took an exponential form and broke it down into its component parts. Now let us take the component parts and turn them into an exponential form. The exponent is the number of component parts. For example www is the same as w^3. There are three w's; therefore, the exponent for w is 3.

EXAMPLE 2:

a)   Write $6QQ$ using exponents. $6Q^2$.

b)   Write $4ttwww$ using exponents $4t^2 w^3$ ■

Now that we have a basic understanding of exponents, we can know learn about the order of operations. Order of operations is the order in which mathematical calculations take place. The order of operations is as follows: all work inside parentheses is done first, raising to a power is second, multiplication and division left to right is done third, and addition and subtraction from left to right is done last to complete the order of operations. There are two ways that the order of operations can be expressed. One way is the anagram PEMDAS. P stands for parentheses, E for exponents, M for multiplication, D for division, A for addition, and S for subtraction. I find this way cold and callous. I prefer the phrase Please Excuse My Dear Aunt Sally where the first letter of each word spells PEMDAS. Every time that the order of operations is needed to solve a problem, I say to my students that Aunt Sally is coming for a visit. My students always know to use the order of operations when Aunt Sally visits. I find this lighthearted approach helps students remember the order of operations better than the cold word PEMDAS. The following examples will show both the correct and incorrect ways to solve order of operations problems. A student can verify the correct way by using a calculator. All calculators use the order of operations to solve problems. The incorrect ways are shown to show students how mistakes can be made if the order of operations is not followed correctly.

EXAMPLE 3 Correct:

Evaluate $6 + 3 \bullet 4$
Since there are no parentheses or exponents involved, the first operation is to multiply.
$6 + 3 \bullet 4$ $3 \bullet 4$ is 12   the problem then becomes
$6 + 12$ the last step is to add
18 is correct the solution. ▄

EXAMPLE 3 Incorrect:

Evaluate $6 + 3 \bullet 4$ $6 + 3$ is 9   the problem then becomes
$9 \bullet 4$ the last step is to multiply
36 is the incorrect solution. ▄

EXAMPLE 4 correct:

Evaluate $6 \bullet 4^2$
Since there are no parentheses involved, the first step is to take care of the exponents.
$6 \bullet 4^2$ 4^2 is 16   the problem then becomes
$6 \bullet 16$ the last step is to multiply
96 is the correct solution. ▄

EXAMPLE 4 incorrect:

Evaluate $6 \bullet 4^2$ $6 \bullet 4$ is 24 the problem then becomes
24^2 the last step is to take care of the exponent
576 is the incorrect solution. ▄

EXAMPLE 5 correct:

Evaluate $8 \bullet 7 + 6 \bullet 5$
Since there are no parentheses or exponents the first step is to multiply.
$8 \bullet 7 + 6 \bullet 5$ $8 \bullet 7$ is 56, and $6 \bullet 5$ is 30   the problem then becomes
$56 + 30$ the last step is to add
86 is the correct solution. ▄

EXAMPLE 5 incorrect:

Evaluate $8 \bullet 7 + 6 \bullet 5$ $7 + 6$ is 13   the problem then becomes

8 • 13 • 5   the last step is to multiply
520 is the incorrect solution.

EXAMPLE 6 correct:

Evaluate $(6 + 3) • 4$
Do work inside parentheses first $6 + 3$ is 9   the problem then becomes
9 • 4   the last step is to multiply
36 is the correct solution. ▪

EXAMPLE 6 incorrect:

Evaluate $(6 + 3) • 4$ ignore the parentheses
6 + 3 • 4   3 • 4 is 12   the problem then becomes
6 + 12   the last step is to add
18 is the incorrect solution. ▪

EXAMPLE 7 correct:

Evaluate $6 • 3^2 + 3$
Since there are no parentheses involved, the first step is to take care of the exponents.
$6 • 3^2 + 3$ 3^2 is 9   the problem then becomes
6 • 9 + 3   the next step is to multiply   6 • 9 is 54   the problem then becomes
54 + 3   the last step is to add
57 is the correct solution. ▪

EXAMPLE 7 incorrect:

Evaluate $6 • 3^2 + 3$ 3^2 is 9 the problem then becomes
6 • 9 + 3   the next step is to add   9 + 3 is 12   the problem then becomes
6 • 12   the last step is to multiply
72 is the incorrect solution. ▪

EXAMPLE 8 correct:

Evaluate $5(3 + 4)^2$
Do work inside parentheses first $3 + 4$ is 7   the problem then becomes
$5(7)^2$ the next step is to take care of the exponent 7^2 is 49   the problem then becomes
5 • 49   the last step is to multiply
245 is the correct solution.
To verify on a calculator enter 5, ×, (, 3, +, 4,), x^2, and =. x^2 is the square button on a calculator. ▪

EXAMPLE 8 incorrect:

Evaluate $5(3 + 4)^2$
Do work inside parentheses first   3 + 4 is 7   the problem then becomes
$5(7)^2$ the next step is to multiply $5 \bullet 7$ is 35   the problem then becomes
35^2 the last step is to take care of the exponent
1225 is the incorrect solution. ■

REMINDER: The purpose of the incorrect solutions is to show where common mistakes are made. Students should not accept these incorrect examples as the correct way of doing these types of problems. Students should always follow the order of operations correctly.

The following two examples are extensions of the order of operations. Only the correct way will be shown.

EXAMPLE 9:

Evaluate $(8 \bullet 3^2 + 4) \div 2$
Do work inside parentheses first. Since there is more than one operation inside the parentheses, the order of operations must be followed.
Since there are no parentheses inside the outer parentheses, the first step is to take care of the exponent inside of the parentheses.
$(8 \bullet 3^2 + 4) \div 2$ 3^2 is 9   the problem then becomes
$(8 \bullet 9 + 4) \div 2$ the next step is to multiply inside the parentheses $8 \bullet 9$ is 72 the problem then becomes
$(72 + 4) \div 2$ the next step is to add inside the parentheses $72 + 4$ is 76 the problem then becomes
$76 \div 2$ once all the work inside the parentheses is completed, the parentheses disappear. The last step is to divide.
38 is the solution. ■

EXAMPLE 10:

Evaluate $(8 \bullet (3^2 \bullet 5) + 4) \div 4$
Do work inside parentheses first. Since there is a set of parentheses inside of the outer parentheses, the order of operations must take place in the inner most parentheses first. The first step is to take care of the exponent in the inner most parentheses. 3^2 is 9   the problem then becomes
$(8 \bullet (9 \bullet 5) + 4) \div 4$ the next step is to multiply inside the inner most parentheses $9 \bullet 5$ is 45 the problem then becomes
$(8 \bullet 45 + 4) \div 4$ notice: once all the work inside the inner most parentheses is completed, the inner most parentheses disappear. The next step is to multiply inside the

remaining parentheses 8 • 45 is 360 the problem then becomes

$(360 + 4) \div 4$ the next step is to add inside the remaining parentheses $360 + 4$ is 364 the problem then becomes

$364 \div 4$ notice: once all the work inside the parentheses is completed, the parentheses disappeared. The last step is to divide.

91 is the solution. ▬

  Enter examples 3 through 10, with the exception of example 8, as given into a calculator to verify the correct solutions. The correct calculator procedure for example 8 is given there.

  Let us now look at a problem that may or may not have the correct answer. Parentheses could be used to correct the problem.

EXAMPLE 11:

$1 + 4 • 3 + 2 = 17$

Check to see if the answer is correct by using order of operations.

$1 + 4 • 3 + 2 = 17$ multiply first.
$1 + 12 + 2 = 17$ add
$15 \neq 17$

Use parentheses to correct the problem, and then check using order of operations.

$(1 + 4) • 3 + 2 = 17$ do inside parentheses first.
$5 • 3 + 2 = 17$ multiply
$15 + 2 = 17$ add
$17 = 17$ ▬

  I could have placed parentheses around $3 + 2$, but that would have led to a different result. It is left to the reader to place parentheses around $3 + 2$ to see the different result. Example 11 is to get readers to think how different placement of parentheses can change results.

PROBLEMS:

Write the following expressions in exponential form.

1) (2)(2)(2)(2)   2) (3)(3)(3)(3)(3)

3) *jjjjj*   4) *5ddd*

5) *ffffhh* 6) *5kkkklll*

7) *7svvvvv*

Solve the following problems by order of operations.

8) $4 + (5)(9)$ 9) $(8 + 3)(7)$

10) $16 - 9 \div 3$ 11) $(10 - 6) \div 2$

12) $(4)(3) + (5)(5)$ 13) $(6)(5 + 8)(8)$

14) $(8)(2^2)$ 15) $(4 \bullet 6)^2$

16) $(2)(4^2) - 3$ 17) $(5)(6^3 - 9)$

18) $(2)(3^4) - (7)(3)$ 19) $(4 \bullet 6)^2 - (5)(6)$

20) $(7)(5 + 9)^2$ 21) $(8 \bullet 6 + 9)^2$

22) $(5)(3 + 4)^2 - 11$ 23) $(2 \bullet 3 + 4)^2 - 15$

Use parentheses to correct the following problems.

24) $5 + 3 \bullet 2 - 4 + 1 = 13$ 25) $4^2 - 2 + 3 \bullet 3 = 23$

SECTION 1.3
PROPERTIES OF ADDITION AND MULTIPLICATION

There are two properties that deal with addition and multiplication as separate operators. There is one property that combines addition and multiplication. We will discuss each property from the least important to the most important.

The associative property is the least important of the three properties in a beginning Algebra course. The associative property will not be seen nor used again after this section. The associative property allows you to group in any manner. For instance, an addition problem, or a multiplication problem with at least three terms can be grouped in different manners, and the results would be the same.

ASSOCIATIVE PROPERTY

$A + (B + C) = (A + B) + C$ for all real numbers
$A (BC) = (AB)C$ for all real numbers

EXAMPLE 1:

a)   Verify $5 + (6 + 7) = (5 + 6) + 7$ remember order of operations must be followed
      Do inside parentheses first. $6 + 7 = 13$ on the left, and $5 + 6 = 11$ on the right. The
      problem then becomes
      $5 + 13 = 11 + 7$ The last step is to add both sides.
      $18 = 18$

b)   Verify $5 \bullet (6 \bullet 7) = (5 \bullet 6) \bullet 7$ remember order of operations must be followed
      Do inside parentheses first. $6 \bullet 7 = 42$ on the left and $5 \bullet 6 = 30$ on the right. The
      problem then becomes
      $5 \bullet 42 = 30 \bullet 7$ The last step is to multiply both sides.
      $210 = 210$ ■

From the above examples, it can be observed that the associative property is nothing more than moving the parentheses.

The commutative property is the next important property. The commutative property allows a changing of the order of an addition problem, or the order of a multiplication problem. Signs should not be changed when using the commutative property. Incorrect results will occur if there are sign changes in an addition problem. Incorrect results may occur if there are sign changes in multiplication. Example 3 will show what happens when changes happen. The commutative property will be a useful tool in Chapter 6.

COMMUTATIVE PROPERTY

A + B = B + A for all real numbers.
AB = BA for all real numbers.

EXAMPLE 2:

a) Verify 4 + 8 = 8 + 4 add both sides
    12 = 12

b) Verify 3 • 7 = 7 • 3 multiply both sides
    21 = 21 ■

EXAMPLE 3:

a) Verify or disprove 8 - 4 = 4 - 8 do the indicated operation on both sides
    4 ≠ - 4

b) Verify or disprove 23 - 23 = 23 - 23 do the indicated operation on both sides
    0 = 0

c) Verify or disprove (- 3) • 7 = (- 7) • 3 multiply both sides.
    - 21 = - 21

d) Verify or disprove (3)(7) = (- 7)(- 3) multiply both sides
    21 = 21

e) Verify or disprove 3 • 7 = 7 • (- 3) multiply both sides
    21 ≠ - 21 ■

Notice that the sign changes in examples 3c and 3d had no bearing on the result, but in examples 3a and 3e the results did change. Example 3b shows that the commutative property only works in subtraction when any number subtracts itself. Negative numbers will be explained in Chapter 2. Students should come back to this page after studying Chapter 2 if they do not understand the results of Example 3. It is not a good idea to change signs under any circumstances. Sign changes will be discussed in later sections if a change is needed.

The most important property is the distributive property. This property combines addition with multiplication. This property will be used throughout most of this book. Example 4 will show algebraic use of the distributive property. Example 5 will show the distributive property as an alternative to the order of operations.

DISTRIBUTIVE PROPERTY (MULTIPLY INTO PARENTHESES)

A (B + C) = AB + BC for all real numbers

EXAMPLE 4:

a)   8(r + u) = 8r + 8u

b)   4(5y + 7) = 20y + 28 ■

Multiplication between numbers and unknowns will be explained in Section 1.5. Students should come back to this page after studying Section 1.5 if they do not understand the results of Example 4.

EXAMPLE 5:

Verify by using the distributive property on the left, and order of operations on the right. 6(4 + 5) = 6(4 + 5) using the distributive property on the left the problem becomes 6 • 4 + 6 • 5 = 6(4 + 5) using the order of operations on both sides the problem becomes 24 + 30 = 6 • 9 add on the left and multiply on the right 54 = 54 ■

Someone may wonder why these properties are only valid for addition and multiplication. Example 3a showed why subtraction cannot be used with the commutative property. Subtraction can be combined with the distributive property as example 6 will show. Example 7 will show that division cannot use the commutative property except in one case.

EXAMPLE 6:

a)   6(p - s) = 6p - 6s

b)   Verify by using the distributive property on the left and the order of operations on the right 4(5 - 3) = 4(5 - 3) using the distributive property on the left the problem becomes
20 - 12 = 4(5 - 3) using the order of operations on the right the problem becomes
20 -12 = 4 • 2 subtract on the left multiply on the right
8 = 8 ■

EXAMPLE 7:

a)   Verify 1 ÷ 1 = 1 ÷ 1 divide both sides
1 = 1

b)   Verify or disprove $4 \div 2 = 2 \div 4$

$2 \neq 0.5$ ■

Example 7a is one possible example where the commutative property works in division. Formally, the commutative property works in division when any number is divided by itself.

PROBLEMS:

Identify the properties illustrated.

1) $2 + 6 = 6 + 2$ 2) $3 \bullet (4 \bullet 6) = (3 \bullet 4) \bullet 6$

3) $11 \bullet 6 = 6 \bullet 11$ 4) $4 + (6 + 7) = (4 + 6) + 7$

Verify the following.

5) $3 + (4 + 5) = (3 + 4) + 5$ 6) $2 \bullet (3 \bullet 6) = (2 \bullet 3) \bullet 6$

Verify by using the distributive property on the left and the order of operations on the right.

7) $2 \bullet (4 + 5) = 2 \bullet (4 + 5)$ 8) $3 \bullet (5 - 2) = 3 \bullet (5 - 2)$

9)   Give an example where the Commutative Property works under subtraction.

10)   Give an example where the Commutative Property works under division.

SECTION 1.4
ADDITION AND SUBTRACTION OF ALGEBRAIC EXPRESSIONS

This section deals with adding and subtracting like and unlike terms. A term can be in the form of a number, a single unknown, a product of unknowns, or an unknown(s) preceded by a number. <u>A number is called a coefficient if the number precedes an unknown.</u> <u>Terms that are separated by the symbols of addition or subtraction are called expressions.</u> Example 1 shows some common terms and expressions.

EXAMPLE 1:

a)   6   b) q c) fw

d) $7a$ e) $8b^2 + 9b$ f) $x - y$ ■

Notice that Example 1e has terms that are separated by a plus sign, and Example 1f has terms that are separated by a minus sign. In both examples, the terms are unlike, and cannot be added or subtracted. For terms to be like, the unknowns must match. The unknown's exponents must also match for the terms to be added. In Example 1e, the unknowns match, but the exponents do not. $8b^2 + 9b$ cannot be added because of the unmatched exponents.

Like terms can be added or subtracted. Example 2 will show common like terms. Example 3 will show the addition and subtraction of like terms. Example 4 will show combinations of unlike and like terms.

EXAMPLE 2:

a)   3, 5 all real numbers are considered like

b) $5b, 3b$

c) $9x^2yz^3, 4x^2yz^3, 13x^2yz^3$ ■

EXAMPLE 3:

a) $3 + 5 = 8$

b) $6c + 4c = 10c$ just add the coefficients, and keep the unknown.

c) $7rs^3 - 2rs^3 = 5rs^3$ just subtract the coefficients, and keep the unknowns, do not touch the exponents. ■

EXAMPLE 4:

a) $8y - 4y + 6z = 4y + 6z$ only add or subtract like terms

b) $7u + 8x + 4u - 5x = 11u - 3x$ ■

      The last example will deal with the concept of an invisible 1. For every unknown or product of unknowns that do not have a numerical coefficient, an invisible 1 is the coefficient. For example, x by itself means that there is $1x$. In algebra, the 1 need not be written, therefore; the 1 is invisible. The invisible 1 is used throughout this book, and may appear in other forms besides a coefficient.

EXAMPLE 5:

a) $5a + 7a + a = 13a$

b) $3x^3 + 5x^3 - x^3 = 7x^3$ ■

      A common error is to forget to add or subtract the invisible 1.

PROBLEMS:

Combine like terms.

1) $2c + 6c$

2) $4g^3 + 11g^3$

3) $12mno + 5mno$

4) $6k^2 - 4k^2$

5) $2c^3 - 2c^3$

6) $15p^2 - 14p^2$

7) $17r^2s - 7r^2s$

8) $23a^2 - 2a^2 + 3a^2$

9) $4b - 2b + 3c$

10) $4g + 2f - g - f$

SECTION 1.5
MULTIPLICATION AND DIVISION OF ALGEBRAIC EXPRESSIONS

We have seen how a number and an unknown represented multiplication back in Section 1.1. We will now see how these types of expressions are multiplied. Example 1 will show simple multiplications, and Example 2 will make use of the first law of exponents. The first law of exponents states that the exponents must be added when the bases are the same.

EXAMPLE 1:

a) $2a \bullet 3b = 6ab$ just multiply the coefficients

b) $4c^2 \bullet 5d^3 = 20c^2d^3$ just multiply the coefficients, and leave the exponents alone ■

In Example 1, notice that the unknowns are not touched.

> **LAW #1 OF EXPONENTS**
>
> $x^m x^n = x^{m+n}$ when the base is the same, just add the exponents

EXAMPLE 2:

a) $g^4 g^6 = g^{4+6} = g^{10}$

b) $h^3 h^3 h^3 = h^{3+3+3} = h^9$ ■

We will now see how combinations of coefficients and unknowns are multiplied together.

EXAMPLE 3:

a) $4b^2 \bullet 5b^5 = 20b^7$ multiply the coefficients, add the exponents on the matching unknowns

b) $6c^3 \bullet 7d^3 = 42c^3d^3$ just multiply the coefficients, unknowns do not match, do not add exponents

c) $8g \bullet 9k \bullet 10g^3 \bullet 11k^3 = 7920g^4k^4$ just multiply the coefficients, add the exponents on the matching unknowns. ■

Example 3c shows another case for the invisible 1. An invisible 1 is used as an exponent when an exponent is not given. Raising an unknown or number to the first power is the same as the unknown or the number itself. A common error is to forget to add the invisible exponents.

What happens to coefficients and exponents when the operation is division instead of multiplication? Example 4 will use the second law of exponents, and Example 5 will show how coefficients are used in combination with the second law of exponents. The second law of exponents states that the exponents must be subtracted when the bases are the same.

LAW #2 OF EXPONENTS

a) $x^m \div x^n = x^{m-n}$ for $m > n$ when the base is the same, just subtract the exponents

b) $x^m \div x^n = 1$ for $m = n$ when the base is the same, and the exponents are the same the result is 1. Remember from earlier math courses, anything divided by itself is 1. We will discuss Law 2b more formally, and introduce Law 2c in Chapter 10.

EXAMPLE 4:

a) $y^7 \div y^3 = y^4$ just subtract exponents

b) $c^4 d^8 \div c^3 d^4 = cd^4$ reminder, 1 does not have to be written as an exponent

c) $g^3 \div g^3 = 1$ Law 2b of exponents ■

EXAMPLE 5:

a) $9y^7 \div 3y^3 = 3y^4$ just divide coefficients and subtract the exponents on matching unknowns

b) $24c^4 d^8 \div 2c^3 d^4 = 12cd^4$ just divide coefficients and subtract the exponents on matching unknowns, reminder, 1 does not have to be written as an exponent

c) $23g^3 \div 23g^3 = 1$ anything divided by itself is 1 ■

It was stated, in Section 1.1, that fractions represent division. This representation will be introduced in Chapter 6. This section only dealt with simple divisions, and not the more complicated ones that will be seen later.

PROBLEMS:

Multiply.

1) $a^2 \bullet a^4$

2) $3 \bullet 3^3$

3) $d^5 \bullet d$

4) $g^{11} \bullet g^5$

5) $k^4 l \bullet k^3 l^2$

6) $c^4 \bullet c \bullet c^5$

7) $a^2 \bullet a^3 \bullet a^5$

8) $o^6 p \bullet o^5 p^7 \bullet o^4 p^8$

9) $2c^4 \bullet 3c^5$

10) $a^3 \bullet 4a^6$

11) $6c^4 d \bullet 5c^2 d^2$

12) $3b^4 c \bullet 6bc^2$

13) $3b^3 \bullet b^4 \bullet 4b^8$

14) $4d^3 g \bullet 5dg^2 \bullet 3d^6 g$

15) $3b^5 \bullet b^6 \bullet 4b \bullet 5b^7$

16) $6b^4 c \bullet bc^5 \bullet 7b^3 c \bullet 9bc$

Divide.

17) $b^4 \div b^3$

18) $g^{16} \div g^2$

19) $c^{20} \div c^{10}$

20) $k^3 l^2 \div k^2 l$

21) $4l^2 \div 2l$

22) $12b^3 \div 2b^2$

23) $46a^4 b \div 23a^2$

24) $27b^3 g^2 \div 9bg$

SECTION 1.6
EVALUATING ALGEBRAIC EXPRESSIONS

It is now time to use what we have learned in the first five sections. This section's only concern is solving problems with known variables (unknowns). Every problem has two parts to it. The first part is a substitution, and the second is the actual solving of the problem. Substitution is nothing more than trading the given number for the unknown that represents that particular number. For example, if $M = 5$ then 5 replaces every M in an expression.

EXAMPLE 1:

Let $C = 6$ and $D = 3$ for all problems

a)   Evaluate $C + D$ by making the appropriate substitutions.
        $6 + 3$ add
        The result is 9.

b)   Evaluate $C - D$ by making the appropriate substitutions.
        $6 - 3$ subtract
        The result is 3.

c)   Evaluate CD by making the appropriate substitutions.
        $6 \bullet 3$ multiply
        The result is 18.

d)   Evaluate $C \div D$ by making the appropriate substitutions.
        $6 \div 3$ divide
        The result is 2.

e)   Evaluate C^2 by making the appropriate substitutions.
        6^2 raise 6 to the second power
        The result is 36. ■

EXAMPLE 2:

Let $E = 6$, $F = 7$, $G = 8$, $H = 9$, and $X = 4$ for all problems

a)   Evaluate $6F + 8H$ by making the appropriate substitutions.
        $6 \bullet 7 + 8 \bullet 9$ multiply first
        $42 + 72$ add
        The result is 114.

b) Evaluate $3G^2$ by making the appropriate substitutions.
    $3 \bullet 8^2$ raise 8 to the second power
    $3 \bullet 64$ multiply
    The result is 192.

c) Evaluate $5E^4 - 2H^2$ by making the appropriate substitutions.
    $5 \bullet 6^4 - 2 \bullet 9^2$ raise 6 to the fourth power and 9 to the second power
    $5 \bullet 1296 - 2 \bullet 81$ multiply
    $6480 - 162$ subtract
    The result is 6318.

d) Evaluate $4F + 3E \div H$ by making the appropriate substitutions.
    $4 \bullet 7 + 3 \bullet 6 \div 9$ multiply first
    $42 + 18 \div 9$ divide
    $42 + 2$ add
    The result is 44.

e) Evaluate $2EF - 3G \div X$ by making the appropriate substitutions.
    $2 \bullet 6 \bullet 7 - 3 \bullet 8 \div 4$ multiply first
    $84 - 24 \div 4$ divide
    $84 - 6$ subtract
    The result is 78. ■

EXAMPLE 3:

Let $E = 6$, $F = 7$, $G = 8$, $H = 9$, and $K = 4$ for all problems

a) Evaluate $6(F + 8H)$ by making the appropriate substitutions.
    $6(7 + 8 \bullet 9)$ multiply inside parentheses first
    $6(7 + 72)$ add inside parentheses
    $6 \bullet 79$ multiply
    The result is 474.

b) Evaluate $(3G)^2$ by making the appropriate substitutions.
    $(3 \bullet 8)^2$ multiply inside parentheses first
    24^2 raise 24 to the second power
    The result is 576.

c) Evaluate $5(E^4 - 2H^2)$ by making the appropriate substitutions.
    $5(6^4 - 2 \bullet 9^2)$ raise 6 to the fourth power and 9 to the second power inside parentheses
    $5(1296 - 2 \bullet 81)$ multiply inside parentheses
    $5(1296 - 162)$ subtract inside parentheses

5 • 1134 multiply
The result is 5670.

d)   Evaluate $(4F + 3E) \div E$ by making the appropriate substitutions.
(4 • 7 + 3 • 6) ÷ 6 multiply first inside parentheses
(42 + 18) ÷ 6 add inside parentheses
60 ÷ 6 divide
The result is 10.

e)   Evaluate $2(EF - 3G) \div K$ by making the appropriate substitutions.
2(6 • 7 - 3 • 8) ÷ 4 multiply inside parentheses
2(42 - 24) ÷ 4 subtract inside parentheses
2 • 18 ÷ 4 multiply first
36 ÷ 4 divide
The result is 9. ■

Notice that in every example not all the unknowns were used in every problem. It is not necessary to use all given unknowns. Only use the unknowns that are needed.

Substituting values into a known formula is another form of substitution. $D = RT$ is the distance formula. The distance is equal to the rate of speed multiplied by the time traveled. I will show a simple problem where the rate and time are known. Problems where the rate or time needs to be solved will be shown in Chapter 3.

EXAMPLE 4:

A car travels 2 hours down a highway at a constant speed of 55 miles per hour. How far has the car traveled?

The time is 2 hours ($T = 2$), and the rate of speed is 55 miles per hour (R =55).

Multiply the rate and speed together.

$D = 2 • 55 = 110$

The car has traveled 110 miles. ■

Other formulas with their substitutions will be shown later in their appropriate sections.

PROBLEMS:

Evaluate if $m = 21$, $n = 80$, $o = 2$, $p = 8$, and $q = 5$.

1) $7n + o$ 2) $5(m + n)$

3) $2op - q$ 4) $5mn \div o$

5) $4q^2$ 6) $(7o)^2$

7) $(p - q)^2$ 8) $p^2 - q^2$

9) $(o + p)^3$ 10) $q^3 + o^3$

11) $6p + 8m$ 12) $5(o^6 - 2q^2)$

13) $(4m + 3o) \div q$ 14) $o^2 p^2 q^2$

15) $5o^2 p$ 16) $2p^2 q + 3pq^2$

17) $4q^2 - 6o^3$ 18) $p^3 - 7mo$

19) How far will a person drive a car at a constant speed of 50 miles an hour for 5 hours?

20) How far will a person drive a car at a constant speed of 25 miles an hour for 15 minutes? (HINT: 15 minutes is one-fourth of an hour.)

CHAPTER 2
THE REAL NUMBER SYSTEM

SECTION 2.1
AN INTRODUCTION TO THE REAL NUMBER SYSTEM

Whole numbers, fractions, and decimals all exist in the real number system. The real number system can be broken into two groups of numbers. Theses groups are the positive numbers and negative numbers. <u>Positive numbers are greater than zero.</u> <u>Negative numbers are less than zero.</u> Zero is neither positive nor negative. Zero is the midpoint of the real number system. The real number system can be represented on a number line. This line is shown in Diagram 1. Each vertical line represents an increase of ¼ or .25. The arrows at the ends show that the number line continues forever in both directions.

DIAGRAM 1:

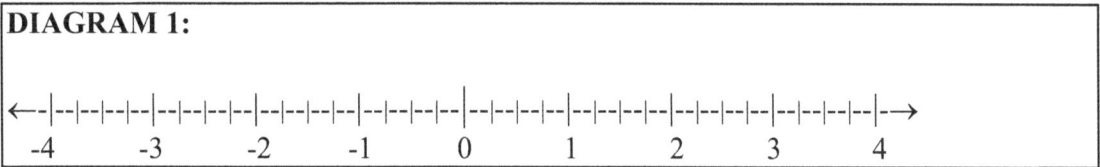

EXAMPLE 1:

a)   2 is a positive number

b)   - 6 is a negative number

A positive sign does not have to be shown if a positive number stands by itself, or if the number is at the beginning of an expression. It is understood that a number without a sign is always positive.

We saw, in Section 1.1, that the basic operations have written words that mean the operations. Positive and negative also have written words that mean a positive number, or a negative number. These word associations are shown in the Word Bank.

WORD BANK	
<u>Positive</u>	<u>Negative</u>
above sea level	below sea level
profit	loss
above zero	below zero

EXAMPLE 2:

a)   26 feet above sea level is the same as 26

b)   21 feet below sea level is the same as - 21

c)   A profit of $39.95 is the same as $39.95

d)   A loss of $34.95 is the same as - $34.95

e)   20 degrees above zero is the same as 20

f)   3 degrees below zero is the same as – 3 ■

Now that we have a basic understanding of positive and negative numbers, let's discuss the concept of absolute value. <u>The absolute value represents the distance a number is from zero.</u> Whether a number is negative or positive does not matter. The absolute value is always a positive number. Distance is always measured as a positive number. For example, I travel 13 miles from my house to work in both directions. I do not say I travel 13 miles in one direction, and - 13 miles in the other direction. I just say I drive 13 miles to and from work. In symbols, $|13|$ read the absolute value of 13 and $|- 13|$ read the absolute value of negative 13. Both $|13|$, and $|- 13|$ equal 13. 13 and - 13 both have a distance of 13 from zero.

EXAMPLE 3:

a) $|5| = 5$

b) $|- 0.26| = 0.26$

c) $|½| = ½$ ■

We learned that all work must be done inside parentheses when we studied the order of operations. All work must also be done inside the absolute value symbol. Think of the absolute value symbol as flattened parentheses. There may be no work inside the absolute value symbols, however; the taking of the absolute value is the work (see Example 3).

EXAMPLE 4:

a) $|14 - 12|$ subtract
         $|2|$ take the absolute value
         The result is 2.

b) $|0.2 + 0.3|$ add
         $|0.5|$ take the absolute value
         The result is 0.5.

c) $|¾ - ¾|$ subtract

$|0|$ take the absolute vale
The result is 0. ■

    Absolute values can be added, subtracted, multiplied, and divided. I will show how absolute values can be used with the basic operations in the appropriate sections. My intent, for this section, was to only give a basic understanding of absolute values.

PROBLEMS:

Translate the following into a signed number.

1)   444 feet (ft.) above sea level

2)   19 ft. below sea level

3)   A $217 loss

4)   A $274 profit

5)   4 degrees above zero

6)   10 degrees below zero

Take the absolute value of the following.

7) $|10|$

8) $|-85|$

9) $|14 - 2|$

10) $|3 + 36|$

11) $|-0.17|$

12) $|0.2|$

13) $|1.1 + 0.2|$

14) $|1.2 - 0.4|$

15) $\left|\dfrac{6}{11}\right|$

16) $\left|\dfrac{-1}{10}\right|$

17) $\left|\dfrac{5}{11} - \dfrac{2}{11}\right|$

18) $\left|\dfrac{3}{13} + \dfrac{4}{13}\right|$

SECTION 2.2
ADDITION AND SUBTRACTION OF REAL NUMBERS

Many authors treat the addition and subtraction of real numbers as two separate sections. I believe that these two operations can be explained in one section. Authors also show these operations with parentheses. Using parentheses to add or subtract real numbers is not the way we write numbers. I will explain the addition and subtraction of real numbers in the way in which we write real numbers. To fully understand the use of parentheses, one must first understand the multiplication of real numbers. The use of parentheses in addition and subtraction will be explained in Section 2.3(Multiplication of real numbers).

Two questions are always asked about adding and subtracting real numbers. First I will answer the addition question then I will answer the subtraction question. When do we add real numbers? To see if you can add real numbers, check to see if the signs are the same in pairs. Addition takes place if the signs are the same in pairs. A pair of positive numbers or a pair of negative numbers can be added together. The order of operations must be used if there is more than one pair of numbers in an expression.

RULE #1:

If a pair of signs are the same, add the numbers, and keep the sign.

EXAMPLE 1:

a) $2 + 8 = 10$ both are positive

b) $- 2.29 - 2.22 = - 4.51$ both are negative ■

When do we subtract real numbers? To see if you can subtract real numbers, check to see if the signs are different in pairs. Subtraction takes place if the signs are different in pairs. A pair of numbers consisting of one positive number and one negative number can be subtracted together.

RULE #2:

The following steps must be taken to subtract real numbers if the signs of the numbers in pairs are different.

Step 1: disregard signs for the time being.
Step 2: subtract using the large number minus the small number
Step 3: mentally take the absolute value of each number
Step 4: in the result, use the sign from the number that had the larger absolute value.

EXAMPLE 2:

a)   9 - 4 = 5   9 had the larger absolute value, therefore the result is positive

b)   5 - 10 = - 5   - 10 had the larger absolute value, therefore the result is negative ■

Before going on to an example that has more than two numbers in it, the concept of subtract from must be explained. This concept has an interesting quality. Problems using this concept may be different than their appearance.

From is the key word in the subtract from concept. The number that comes after from is the number that is written first when translating the sentence into a mathematical expression. The number after from is the stating point. A number is being subtracted from the number after from. **WARNING:** once the sentence is translated into a mathematical expression, the problem may not be a subtraction problem.

EXAMPLE 3:

a)   Subtract 5 from - 7 rewrite
      - 7 - 5 problem is now an addition problem
      The result is - 12

b)   Subtract 12 from 23 rewrite
      23 - 12 problem is a subtraction problem
      The result is 11 because 23 has the larger absolute value.

c)   Subtract 2.54 from 2.32 rewrite
      2.32 - 2.54 problem is a subtraction problem
      The result is - 0.22 because - 2.54 has the larger absolute value ■

Now let us see how the order of operations is applied to the addition of subtraction of real numbers. **REMINDER:** addition or subtraction takes place from left to right.

EXAMPLE 4:

a)   - 3 - 4 - 9 look at first pair of numbers, add
      - 7 - 9 add again
      The result is - 16.

b)   12 + 23 + 2 look at first pair of numbers, add
      35 + 12 add again
      The result is 47. ■

EXAMPLE 5:

a)   7/3 - 5/3 + 2/3 look at first pair of numbers, subtract
2/3 + 2/3   2/3 is positive because 7/3 has the larger absolute value, add
The result is 4/3. An improper fraction is a legal answer in Algebra. There is no
need to change an improper fraction into a mixed number.

b)   - 6.5 + 8.4 - 9.6 look at first pair of numbers, subtract
1.9 - 9.6   1.9 is positive because 8.4 has the larger absolute value, subtract
The result is - 7.7 because - 9.6 has the larger absolute value. ■

Now let us do problems with absolute values.

EXAMPLE 6:

a) $|-3| + |4|$ take the absolute values first
3 + 4 add
The result is 7.

b) $|14| - |7|$ take the absolute values first
14 - 7 subtract
The result is 7.

c) $|8| - |16|$ take the absolute values first
8 - 16 subtract
The result is - 8.

d) $|-3 + 9 - 2|$ do work inside first
$|4|$ take absolute value
The result is 4.

e) $|5 - 3| + |-2 + 6|$ do work inside first
$|2| + |4|$ take absolute values
2 + 4 add
The result is 6. ■

There are two properties associated with the addition and subtraction of real
numbers. The properties are the additive identity property, and the additive inverse
property. The additive inverse property is the more important of the two because it will
be used to solve equations in Chapter 3.

The additive Identity property deals with the addition and subtraction of 0.
Nothing changes when 0 is added to or subtracted from a number.

ADDITIVE IDENTITY PROPERTY

Let n be any number
$n + 0 = 0 + n = n$
$n - 0 = -0 + n = n$

Notice that the commutative property was used to show that the order in which 0 is added does not matter.

EXAMPLE 7:

a) $7 + 0 = 7$

b) $0 - 0.9 = -0.9$ The result is negative because -0.9 has the larger absolute value.

The additive inverse property deals with opposites. When opposites are combined the result is 0. <u>The opposite of a number is the same number but with the opposite sign.</u> 1 and -1 are opposites.

ADDITIVE INVERSE PROPERTY

Let n be any number
$n - n = -n + n = 0$

Notice that the commutative property was used to show that the order in which the opposites are combined does not matter.

EXAMPLE 8:

a) $4 - 4 = 0$

b) $-2.1 + 2.1 = 0$

c) $1/5 - 1/5 = 0$

All examples can be verified with the use of a calculator.

When using negative numbers on a calculator for addition and subtraction problems, the +/- key must be used if a negative number comes first in an addition or in a subtraction problem. The remaining numbers are put into the calculator exactly as they are seen. $-3 + 7 - 5 + 2 =$ is entered into a calculator as follows: 3 +/- key + 7 − 5 + 2 =. The +/- key changes the 3 to a negative 3. For some unknown reason, negative numbers cannot be entered as − 3 in a calculator at the beginning of a problem.

PROBLEMS:

Do the indicated operation.

1) $2 + 4$

2) $21 + 2$

3) $-7 - 2$

4) $3 - 2$

5) $3 - 21$

6) $-2 + 0$

7) $15 + 0$

8) $16 - 16$

9) $-2 - 10 + 2$

10) $2 - 3 - 4 + 5$

11) $6 - 7 + 8 - 9$

12) $-1.1 - 1.2$

13) $-10.2 + 10.4$

14) $-1.1 + 1.7$

15) $1.8 - 1.9 - 2.9$

16) $|-4| + |5|$

17) $|-6| + |6|$

18) $|5| - |-5|$

19) $|6 - 7|$

20) $|-16 + 2|$

21) $|-4 + 5 - 7|$

22) $|3 - 4| + |-7 + 2|$

23) $\dfrac{6}{11} + \dfrac{4}{11}$

24) $-\dfrac{1}{10} - \dfrac{6}{10}$

25) $\dfrac{3}{11} - \dfrac{2}{11}$

26) $-\dfrac{7}{13} - \dfrac{3}{13} + \dfrac{2}{13}$

SECTION 2.3
MULTIPLICATION OF REAL NUMBERS

Signs play an important part in the multiplication of real numbers. Look at pairs of numbers being multiplied to determine if the result is a positive number or a negative number. The result is positive if a pair of multiplied numbers has the same sign. In other words, two positive numbers or two negative numbers multiplied together will result in a positive number. The result is negative when a pair of numbers multiplied together has different signs. A positive number times a negative number results in a negative number.

EXAMPLE 1:

a)　　$6 \bullet 8 = 42$ two positive numbers multiplied together, result is positive

b)　　$(-4)(-9) = 36$ two negative numbers multiplied together, result is positive

c)　　$(-4) \bullet 8 = -32$ one negative number and one positive number multiplied together, result is negative

d)　　$7 \bullet (-1.5) = -10.5$ one positive number and one negative number multiplied together, result is negative ▪

Now let us apply the order of operations for multiplying more than two numbers.

EXAMPLE 2:

a)　　$(-3)(6)(-8)$ multiply first two numbers first
　　　$(-18)(-8)$ multiply again
　　　The result is 144.

b)　　$(-4)(8)(7)$ multiply first two numbers first
　　　$(-32)(7)$ multiply again
　　　The result is - 224.

c)　　$(-3)(-6)(-8)$ multiply first two numbers first
　　　$(18)(-8)$ multiply again
　　　The result is - 144.

d)　　$5 \bullet 9 \bullet 3$ multiply the first two numbers first
　　　$45 \bullet 3$ multiply again
　　　The result is 135. ▪

Now let us multiply absolute values together.

EXAMPLE 3:

a) $|-2| \bullet |3|$ take absolute values first
   $2 \bullet 3$ multiply
   The result is 6.

b) $|15| \bullet |8|$ take the absolute values first
   $15 \cdot 8$ multiply
   The result is 120.

c) $|-7| \bullet |-8|$ take the absolute values first
   $7 \bullet 8$ multiply
   The result is 56.

d) $|4| \bullet |-4|$ take the absolute values first
   $4 \bullet 4$ multiply
   The result is 16. ■

I mentioned in Section 2.2 that addition and subtraction problems that had parentheses were based on multiplication. To understand this concept you must recall the invisible 1 from Chapter 1. With regard to parentheses, there is always an invisible 1 between a sign and the parentheses. For example + (27) has an invisible 1 between the plus sign and (27), and - (57) has an invisible 1 between the minus sign and - (57). Recall from Chapter 1 that a number in front of parentheses implies multiplication.

EXAMPLE 4:

a) $3 + (+ 8)$ multiply 1 and 8 first
   $3 + 8$ add
   The result is 11.

b) $- 7 - (+ 9)$ multiply - 1 and 9 first
   $- 7 - 9$ add
   The result is - 16.

c) $6 + (- 4)$ multiply 1 and - 4 first
   $6 - 4$ subtract
   The result is 2.

d) $4 - (+ 6)$ multiply - 1 and 6 first
   $4 - 6$ subtract
   The result is - 2. ■

Now let us see what happens when a minus sign is placed in front of parentheses, and inside the parentheses is a negative sign. This idea could look something like - (- 2). There is an invisible 1 between the minus sign and the parentheses. We said earlier that a pair of numbers with the same sign that are multiplied results in a positive number. Two negative numbers multiplied together always results in a positive number. - (-2) = -1(-2) = 2.

EXAMPLE 5:

a)   6 - (- 3) multiply - 1 and - 3 first
       6 + 3 add
       The result is 9.

b)   - 11.7 - (- 4.6) multiply - 1 and - 4.6 first
       - 11 .7 + 4.6 subtract
       The result is - 7.1.

c)   Subtract - 9/5 from - 3/5 rewrite
       - 3/5 - (- 9/5) multiply - 1 and - 9/5 first
       -3/5 + 9/5 subtract
       The result is 6/5. ■

Now let us place a minus sign in front of an absolute value. Recall that an absolute value sign can be treated as flat parentheses.

EXAMPLE 6:

a)   - |- 3| This states that - 1 is going to multiply the absolute value of - 3, take absolute value first
       - 1 • 3 multiply
       The result is - 3.

b) |17| - |- 3| take the absolute values first
       17 - 1 • 3 multiply first
       17 - 3 add
       The result is 14.

c) |- 27| - |- 35| take the absolute values first
       27 - 1 • 35 multiply first
       27 - 35 subtract
       The result is - 8. ■

There are three properties associated with the multiplication of real numbers. The properties are the multiplicative identity property, the multiplicative inverse property and

42

the multiplicative property of zero. The multiplicative inverse property is the most important of the three because it will be used to solve equations in Chapter 3.

The multiplicative identity property deals with multiplication by a positive 1. Nothing changes when a positive 1 is multiplied to a number or an unknown.

MULTIPLICATIVE IDENTITY PROPERTY

Let x be any number
$x \bullet 1 = 1 \bullet x = x$.

Notice that the commutative property was used to show that the order in which multiplication takes place does not matter.

EXAMPLE 7:

a) $1(- 8) = - 8$

b) $1.2 \bullet 1 = 1.2$

c) $- 2/5 \bullet 1 = - 2/5$ ■

The multiplicative inverse property states that when a number is multiplied by its reciprocal, the result is a positive 1. Recall that any whole number can be written as a fraction with the denominator being 1. 5 is the same as 5/1. Just turn a fraction upside down to find the reciprocal. Do not change the signs when taking the reciprocal. Recall that the result is positive when two positive or two negative numbers are multiplied together.

MULTIPLICATIVE INVERSE PROPERTY

Let z be any number
$(z/1)(1/z) = z/z = 1$

Recall from fractional arithmetic that when fractions are multiplied, the numerators are multiplied together, and so are the denominators.

EXAMPLE 8:

a)   Find the multiplicative inverse of 3, and show that 3 times its inverse equals 1.
       The multiplicative inverse of 3 is 1/3.
       $3/1 \bullet 1/3 = 3/3 = 1$

b)   Find the multiplicative inverse of - 2/5, and show that - 2/5 times its inverse equals 1.
The multiplicative inverse of - 2/5 is - 5/2.
(- 2/5)(- 5/2) = 10/10 = 1 ■

The multiplicative property of zero only concerns itself with multiplication by 0. 0 multiplied to anything results in 0.

MULTIPLICATIVE PROPERTY OF ZERO

Let m be any number
m • 0 = 0m = 0.

Notice that the commutative property was used to show that the order in which multiplication takes place does not matter.

EXAMPLE 9:

a)   7 • 0 = 0

b)   0(- 0.4) = 0

c)   3/11 • 0 = 0 ■

Most examples can be verified with the use of a calculator. Any example that has absolute values in it can only be verified if a calculator has the ABS function.

When using negative numbers on a calculator for multiplication and division problems, the +/- key must be used for all negative numbers when multiplying or dividing. - 3 * 7 = is entered into a calculator as follows: 3, +/- key, *, 7, =. –21/-7 is entered into a calculator as follows: 21, +/- key, /, 7, +/- key =. The +/- key changes any positive number into a negative number. The opposite is also true. The +/- key changes any negative number into a positive number.

PROBLEMS:

Do the indicated operations. Reduce fractions if possible.

1)   2 • 27   2)   4(-25)

3)   -7 • 2   4)   1.31(-2)

5)   (-3) (-2)   6)   (-3) (-20)

7) (-2.08) (-16)

8) 0(-15)

9) 27 • 0

10) (-1.01) (0)

11) $6\left(\dfrac{-4}{11}\right)$

12) $-\dfrac{1}{10} • 6$

13) $(-3)\left(-\dfrac{2}{11}\right)$

14) $-\dfrac{7}{13} • 0$

15) $\left(-\dfrac{6}{11}\right)\left(-\dfrac{11}{6}\right)$

16) $\dfrac{3}{11}\left(-\dfrac{11}{3}\right)$

17) (-2) (3) (-4)

18) (6) (-3) (4)

19) (-2) (-9) (-4)

20) (-4) (-2) (5)

21) (-6) (-49) (0)

22) $|-2| • |8|$

23) $|14| • |3|$

24) $|-2| • |-3|$

25) 3 + (+ 6)

26) - 4 - (+ 2)

27) 3 + (- 4)

28) 8 - (+ 2)

29) 3 - (- 7)

30) - 11.8 - (- 4.3)

31) $-|-2|$

32) $|14| - |-3|$

SECTION 2.4
DIVISION OF REAL NUMBERS

The division of real numbers is similar to that of multiplication with respect to the sign rules. The answer is positive if two numbers of the same sign are being divided. The answer is negative if two numbers with different signs are being divided.

EXAMPLE 1:

a) $42 \div 6 = 7$ positive divided by positive equals positive

b) $-28 \div -4 = 7$ negative divided by negative equals positive

c) $24 \div -6 = -4$ positive divided by negative equals negative

d) $-55 \div 5 = -11$ negative divided by positive equals negative ■

To verify the answers in Example 1, a calculator can be used, or just multiply the answer with the number to the right of the division symbol. From Example 1a, $7 \bullet 6 = 42$. This multiplication verification will be useful in the discussion of the division definition of zero.

The Division definition of zero has two parts. The first part shows zero divided by anything is zero. The second part shows that division by zero does not exist. The multiplication verification will be used to explain both parts.

DIVISION DEFINITION OF ZERO

a)   Let x be any number
$0 \div x = 0$

Verification $0x = 0$. We know this is a true statement from the Multiplication Property of Zero.

b)   Let y be any number other than 0, and z be the solution other than 0
$y \div 0 = z$

Verification $0z = y$ cannot happen, $0z = y$ states that some number times 0 equals another number. This explanation contradicts the Multiplication Property of Zero, therefore; division by zero is undefined.

EXAMPLE 2:

a) $0 \div -9 = 0$

b) $0 \div 1/3 = 0$

The result is the same whether zero is divided by a positive or negative number.

c) $30 \div 0$ is undefined.

d) $-3.6 \div 0$ is undefined.

The result is the same whether a positive number or a negative number is divided by zero. ■

The mathematical curious may ask about $0 \div 0$. $0 \div 0$ has applications in Calculus. This concept will be discussed in Calculus, and is inappropriate for anyone studying simple algebra.

Now let us see what happens when we divide by 1, and -1.

EXAMPLE 3:

a) $14 \div 1 = 14$ positive divided by positive equals positive

b) $-21 \div 1 = -21$ negative divided by positive equals negative

c) $22 \div -1 = -22$ positive divided by negative equals negative

d) $-31 \div -1 = 31$ negative divided by negative equals positive ■

Notice that there were no sign changes when numbers were divided by 1. There was a sign change when numbers were divided by -1. This information will be useful in chapter 3 for certain problems.

Now let us divide absolute values together.

EXAMPLE 4:

a) $|-6| \div |3|$ take absolute values first
        $6 \div 3$ divide
        The result is 2.

b) $|16| \div |8|$ take the absolute values first
16 ÷ 8 divide
The result is 2.

c) $|- 72| \div |- 8|$ take the absolute values first
72 ÷ 8 divide
The result is 9.

d) $|4| \div |- 4|$ take the absolute values first
4 ÷ 4 divide
The result is 1. ■

PROBLEMS:

Divide.

1)   -49 ÷ -7   2)   60 ÷ 2

3)   14 ÷ -2   4)   -99 ÷ 3

5)   0 ÷ -4   6)   -2 ÷ -1

7)   11 ÷ 0   8)   -73 ÷ 1

9)   -118 ÷ -59   10)   -15.3 ÷ 7.5

11) $|-4| \div |2|$ 12) $|58| \div |2|$

13) $|-30| \div |-2|$ 14) $|2| \div |-2|$

SECTION 2.5
EVALUATING EXPRESSIONS WITH REAL NUMBERS

This section is almost identical to Section 1.6. In this section, the substitution problems will have both positive and negative numbers. Positive numbers are substituted just like before. The substitution of negative numbers is a little different. Place negative numbers inside parentheses after the substitution provided that the substitution is not at the beginning of an expression. Once a negative number is substituted, it can be clearly seen what type of operation takes place with the negative number. The examples in this section will be broken into 3 parts. Part 1 will have one operation, part 2 will have order of operations without parentheses, and part 3 will have order of operations with parentheses. An example with new notation and an example regarding exponents with negative numbers will occur after Example 3.

EXAMPLE 1:

Let $C = -6$ and $D = 3$ for all problems.

a) Evaluate $C + D$ by making the appropriate substitutions.
    $-6 + 3$ subtract
    The result is -3.

b) Evaluate $C - D$ by making the appropriate substitutions.
    $-6 - 3$ add
    The result is -9.

c) Evaluate CD by making the appropriate substitutions.
    $-6 \bullet 3$ multiply
    The result is -18.

d) Evaluate $C \div D$ by making the appropriate substitutions.
    $-6 \div 3$ divide
    The result is -2. ■

EXAMPLE 2:

Let $E = -6$, $F = -7$, $G = -8$, $H = 9$, and $I = 4$ for all problems.

Evaluate $6F + 8H$ by making the appropriate substitutions.
$6(-7) + 8 \bullet 9$ multiply first
$-42 + 72$ subtract
The result is 30. ■

EXAMPLE 3:

Let $E = -6$, $F = -7$, $G = -8$, $H = 9$, and $I = 4$ for all problems.

a)   Evaluate $5E - 2H^2$ by making the appropriate substitutions.
   $5(-6) - 2 \bullet 9^2$ raise 9 to the second power
   $5(-6) - 2 \bullet 81$ multiply
   $-30 - 162$ add
   The result is -192.

b)   Evaluate $4F + 3E \div H$ by making the appropriate substitutions.
   $4(-7) + 3(-6) \div 9$ multiply first
   $-28 - 18 \div 9$ divide
   $-28 - 2$ add
   The result is -30.

c)   Evaluate $2EF - 3G \div I$ by making the appropriate substitutions.
   $2(-6) \bullet 7 - 3(-8) \div 4$ multiply first
   $-84 + 24 \div 4$ divide
   $-84 + 6$ subtract
   The result is -78. ▪

EXAMPLE 4:

Let $E = 6$, $F = -7$, $G = -8$, $H = 9$, and $I = -4$ for all problems.

a)   Evaluate $6(F + 8H)$ by making the appropriate substitutions.
   $6(-7 + 8 \bullet 9)$ multiply inside parentheses first
   $6(-7 + 72)$ subtract inside parentheses
   $6 \bullet 65$ multiply
   The result is 390.

b)   Evaluate $5(E - H - G)$ by making the appropriate substitutions.
   $5(6 - 9 - (-8))$ multiply inside parentheses first, remember invisible 1
   $5(6 - 9 + 8)$ subtract first two numbers inside parentheses first
   $5(-3 + 8)$ subtract inside parentheses
   $5 \bullet 5$ multiply
   The result is 25.

c)   Evaluate $(6F + 3E) \div E$ by making the appropriate substitutions.
   $(6(-7) + 3 \bullet 6) \div 6$ multiply first inside parentheses
   $(-36 + 18) \div 6$ subtract inside parentheses
   $-18 \div 6$ divide

The result is - 3.

d)   Evaluate $2(EF - 3G) \div I$ by making the appropriate substitutions.
      $2(6(- 7) - 3(- 8)) \div (- 4)$ multiply inside parentheses
      $2(- 42 + 24) \div (- 4)$ subtract inside parentheses
      $2(- 18) \div (- 4)$ multiply first
      $- 36 \div (- 4)$ divide
      The result is 9. ■

   Addition and subtraction do not have any new notations. Multiplication, division, and exponentials have new notations. The new notation for multiplication is *. The new division symbol is /. Exponentials now have the ^. The old symbols, from Chapter 1, are still valid. These new symbols are widely used with computer applications such as spreadsheets.

EXAMPLE 5:

Let $A = 3$, $B = - 5$, and $C = 7$ for all problems.

a)   Evaluate $A + B * (- C)$ by making the appropriate substitutions.
      $3 + (- 5) * (- 7)$ multiply
      $3 + 35$ add
      The result is 38.

b)   Evaluate $C ^ 2$ by making the appropriate substitutions.
      $6 ^ 2$ raise 6 to the second power
      The result is 36.

c)   Evaluate $- B - A * C ^ 2$ by making the appropriate substitutions.
      $- (- 5) - 3 * 7 ^ 2$ raise 7 to the second power
      $- (- 5) - 3 * 49$ multiply, remember the invisible 1
      $5 - 147$ subtract
      The result is - 142. ■

   Negative numbers that are raised to a power can be tricky. A negative number that is enclosed in parentheses gets multiplied by itself. For example, $(- 5)^2 = (- 5)(- 5) = 25$. A negative number that is not enclosed in parentheses is tricky. Only the number gets multiplied by itself. The negative sign is untouched. For example, $- 5^2 = - 5 * 5 = -25$. $-5^2 = - 5 * 5 = -25$ is the same as $- 1 * 5 * 5 = -25$.

EXAMPLE 6:

Let $G = -8$ and $H = 6$ for all problems

a)   Evaluate $3G^2$ by making the appropriate substitutions.
       $3(-8)^2$ raise -8 to the second power
       $3 \bullet 64$ multiply
       The result is 192.

b)   Evaluate $(3G)^2$ by making the appropriate substitutions.
       $(3(-8))^2$ multiply inside parentheses first
       $(-24)^2$ raise -24 to the second power
       The result is 576.

c)   Evaluate $H^2 - G^2$ by making the appropriate substitutions.
       $6^2 - (-8)^2$ raise 6 and -8 to the second power
       $36 - 64$ subtract
       The result is -28.

d) $-H^2 - G^2$ by making the appropriate substitutions.
       $-6^2 - (-8)^2$ raise 6 and -8 to the second power
       $-36 - 64$ add
       The result is -100. ■

     As before, all results can be verified with the aid of a calculator.

PROBLEMS:

Do the indicated operations using $b = -2$, $c = 4$, $d = -5$, and $g = 7$.

1) $5d - 2c$ 2) $3c + 4d$

3) $-c^2 + c$ 4) $5b^2$

5) $d^2 - 6g$ 6) $7b^2 + 8c^2$

7) $9(b + c)$ 8) $2(7b - g)$

9) $b(c + 8d)$ 10) $10f \div d$

11) $g^2 - c^2$ 12) $(g - c)^2$

13) $(g - c)(g + c)$ 14) $g^3 - c^3$

52

CHAPTER 3
EQUATIONS AND INEQUALITIES

SECTION 3.1
SOLVING ALGEBRAIC EQUATIONS USING THE ADDITIVE INVERSE

An equation shows that two expressions are equal. In other words, both sides of the equal sign must equal. We have seen equations with regard to the commutative and associative properties. One must understand the definition of an equation before solving an algebraic equation because it is the solution of the algebraic equation that makes the equation. An algebraic equation has at least one unknown in it.

$x + 1 = 2$ is a simple algebraic equation. This section deals with using the additive inverse to solve algebraic equations.

There are two goals that must be achieved is solving an algebraic equation. The unknown must be on the left side of the equal sign, and real numbers must be on the right side. Let us examine $x + 1 = 2$ more closely. The unknown is on the left side of the equal sign, therefore; one goal has been achieved. There are real numbers on both sides of the equal sign. The additive inverse must be used to achieve the second goal. The additive inverse is applied on the left side of the equal sign. The additive inverse of + 1 is - 1. Example 1 will show how the additive inverse is used.

EXAMPLE 1:

Solve for x if $x + 1 = 2$

x is on the left, therefore; goal 1 is achieved.

Numbers are on both sides of equal sign, take additive inverse of number on left side. The additive inverse of + 1 is - 1.
Rewrite algebraic equation as
$x + 1 = 2$
$\quad - 1 \quad - 1$
combine down

$x = 1$ which is a solution. To find out if $x = 1$ is the solution; replace the x with 1 in the original equation and check.
$1 + 1 = 2$ add on left
$2 = 2$ checked. $x = 1$ is the solution to the algebraic equation x + 1 = 2. ∎

Notice that the additive inverse was used on both sides of the equal sign. The additive inverse must be used on both sides of the equal sign to balance the algebraic equation. By balancing the algebraic equation, the correct solution will occur. An unbalanced equation could have disastrous results in a chemistry class. The same is true in Algebra. An unbalanced algebraic equation will yield incorrect solutions.

EXAMPLE 2:

Solve for x if $x - 5 = 11$

x is on the left, therefore; goal 1 is achieved.

Numbers are on both sides of equal sign, take additive inverse of number on left side.
The additive inverse of - 5 is + 5.
Rewrite algebraic equation as
$x - 5 = 11$
$+ 5 \quad + 5$
combine down

$x = 16$ which is a solution. To find out if $x = 16$ is the solution, replace the x with 16 in the original equation and check.
$16 - 5 = 11$ subtract on left
$11 = 11$ checked. $x = 16$ is the solution to the algebraic equation $x - 5 = 11$. ■

EXAMPLE 3:

Solve for x if $x + 4 = 8$

x is on the left, therefore; goal 1 is achieved.

Numbers are on both sides of equal sign, take additive inverse of number on left side.
The additive inverse of + 4 is - 4.
Rewrite algebraic equation as
$x + 4 = 8$
$- 4 \quad - 4$
combine down

$x = 4$ which is a solution. To find out if $x = 4$ is the solution, replace the x with 4 in the original equation and check.
$4 + 4 = 8$ add on left
$8 = 8$ checked. $x = 4$ is the solution to the algebraic equation x + 4 = 8. ■

EXAMPLE 4:

Solve for x if $7x = 6x + 10$

10 is on the right, therefore; goal 2 is achieved.

Unknowns on both sides of equal sign, take additive inverse of unknown on right side.
The additive inverse of $6x$ is - $6x$.
Rewrite algebraic equation as

$7x = 6x + 10$
$-6x \quad - 6x$
combine down

$x = 10$ which is a solution. To find out if $x = 10$ is the solution, replace the x with 10 in the original equation and check.
$7(10) = 6(10) + 10$ multiply first
$70 = 60 + 10$ add on right
$70 = 70$ checked. $x = 10$ is the solution to the algebraic equation $7x = 6x + 10$. ■

EXAMPLE 5:

Solve for x if $9x - 10 = 8x + 6$

Numbers and unknowns are on both sides of equal sign; take additive inverse of number on left side, and additive inverse of unknown on right side.
The additive inverse of - 10 is + 10, and the additive inverse of $8x$ is - $8x$.
Rewrite algebraic equation as
$9x - 10 = 8x + 6$
$-8x + 10 - 8x + 10$
combine down

$x = 16$ which is a solution. To find out if $x = 16$ is the solution, replace the x with 16 in the original equation and check.
$9(16) - 10 = 8(16) + 6$ multiply first
$144 - 10 = 128 + 6$ subtract on left   add on right
$134 = 134$ checked. $x = 16$ is the solution to the algebraic equation $9x - 10 = 8x + 6$. ■

A check was performed in the above examples to verify that the solution is the correct one. A false statement would have occurred if the found solution was an incorrect one.

The next two examples will show how to solve equations when there is work to be done on both sides of the equations. Work can be of the form of combining like terms, or using the distributive property.

EXAMPLE 6:

Solve for x if $7 + 8x - 4 = 2x - 5 + 5x$

Like terms can be combined on both sides. The Algebraic Equation becomes
$3 + 8x = 7x - 5$

Numbers and unknowns are on both sides of equal sign; take additive inverse of number on left side, and additive inverse of unknown on right side.

The additive inverse of 3 is - 3, and the additive inverse of $7x$ is - $7x$.
Rewrite algebraic equation as
$3 + 8x = 7x - 5$
$- 3 - 7x - 7x - 3$
combine down

$x = - 8$ which is a solution. To find out if $x = - 8$ is the solution, replace the x with - 8 in the original equation and check.
$7 + 8(- 8) - 4 = 2(-8) - 5 + 5(- 8)$
$7 - 64 - 4 = - 16 - 5 - 40$ subtract on left add on right
$- 57 - 4 = - 21 - 40$ add
$- 61 = - 61$ checked. $x = - 8$ is the solution to the algebraic equation $7 + 8x - 4 = 2x - 5 + 5x.$ ■

EXAMPLE 7:

Solve for x if $5(5x + 2) = 3(8x - 4) - 5$

Use distributive property on both sides. Algebraic Equation becomes
$25x + 10 = 24x - 12 - 5$ combine like terms on right
$25x + 10 = 24x - 17$

Numbers and unknowns are on both sides of equal sign; take additive inverse of number on left side, and additive inverse of unknown on right side.
The additive inverse of + 10 is - 10, and the additive inverse of $24x$ is - $24x$.
Rewrite algebraic equation as
$25x + 10 = 24x - 17$
$-24x - 10$ $-24x - 10$
combine down

$x = - 27$ which is a solution. To find out if $x = - 27$ is the solution, replace the x with - 27 in the original equation and check.
$5(5(- 27) + 2) = 3(8(- 27) - 4) - 5$ multiply inside parentheses first
$5(- 135 + 2) = 3(- 216 - 4) - 5$ subtract inside parentheses on left, and add inside parentheses on right
$5(- 133) = 3(- 220) - 5$ multiply
$- 665 = - 660 - 5$ add on right
$- 665 = - 665$ checked. $x = - 665$ is the solution to the algebraic equation
$5(5x + 2) = 3(8x - 4) - 5.$ ■

x was used in the above equations to represent the unknown. Most letters in the alphabet can be used to represent the unknown. x is the conventional variable. The letters e and i are not normally used as variables because these two letters are used to represent specific numbers. e represents the base of natural logarithms which is 2.71828182846 and i is the square root of − 1 which is a complex number. Only complex numbers will be discussed in this book. Natural logarithms are discussed in upper level algebra books.

Some equations may have fractions in them. Just solve the equations in the same manner as the examples. Check the solution in the same manner also if a fraction occurs as a solution.

PROBLEMS:

Solve for the unknown.

1) $b + 2 = 58$

2) $a - 2 = 8$

3) $h - 8 = -30$

4) $a + 2 = -4$

5) $7f = 6f + 5$

6) $22u = 21u - 29$

7) $2p = p + \frac{1}{2}$

8) $5m = 4m + \frac{2}{3}$

9) $2a + 3 = a + 4$

10) $6f - 7 = 5f - 8$

11) $2j - 7 = j + 8$

12) $7 + 2j + 8 = 4j + 20 - 3j$

13) $4q + 6 + 7q = 6q + 24 + 4q$

14) $2a - 3 + 4a - 5 + a = 6a + 7$

15) $4(4a + 8) = 15a - 2$

16) $2(4c + 5) = 7(c + 7) + 13$

17) $4(7x - 1) + 3(2 - 5x) = 4(3x + 5)$

18) $7(x - 3) = 4 + 2(3x - 5)$

19) $9(2m - 3) - 4(5 + 3m) = 5(4 + m)$

20) $9(x - 2) = 8 + 4(2x - 1)$

SECTION 3.2
SOLVING ALGEBRAIC EQUATIONS
USING THE MULTIPLICATIVE INVERSE

In the previous section, the additive inverse was used to solve algebraic equations. Every unknown had a coefficient of 1 after the additive inverse was applied. What happens when an unknown has a coefficient other than 1? The multiplicative inverse is used to make the unknown's coefficient equal to 1.

EXAMPLE 1:

Solve for x if $8x = 24$ Number on right side of equation, goal 2 is achieved
        Unknown on left side of equation but has a coefficient other than 1.

Take unknown's coefficient's multiplicative inverse.
The multiplicative inverse of 8 is 1/8.
Multiply both sides by multiplicative inverse
$(1/8)8x = 24(1/8)$
$x = 3$ check solution in original algebraic equation
$8 * 3 = 24$
$24 = 24$ checked $x = 3$ is the solution to the algebraic equation $8x = 24$. ■

Multiplying by 1/8 is the same as dividing by 8.

EXAMPLE 2:

Solve for x if $x/3 = 9$ Number on right side of equation, goal 2 is achieved
        Unknown on left side of equation but has a coefficient other than 1.

Take unknown's coefficient's multiplicative inverse.
$x/3$ is the same as $(1/3)x$
The multiplicative inverse of 1/3 is 3.
Multiply both sides by multiplicative inverse
$3 * x/3 = 9 * 3$
$x = 27$ check solution in original algebraic equation
$27/3 = 9$
$9 = 9$ checked $x = 27$ is the solution to the algebraic equation $x/3 = 9$. ■

Recall from chapter 1 that a fraction is a form of division.

EXAMPLE 3:

Solve for x if $(5/8)x = 10$ Number on right side of equation, goal 2 is achieved
        Unknown on left side of equation but has a coefficient other than 1.
Take unknown's coefficient's multiplicative inverse.

The multiplicative inverse of 5/8 is 8/5.
Multiply both sides by multiplicative inverse
$(8/5)(5/8)x = 10(8/5)$
$x = 16$ check solution in original algebraic equation
$(5/8) * 16 = 10$
$10 = 10$ checked $x = 16$ is the solution to the algebraic equation $(5/8)x = 10$. ■

Recall from fractional arithmetic, during multiplication, denominators can cancel into numerators. For 10(8/5), 5 cancels into 10 two times, and 2 * 8 = 16. For (5/8) * 16, 8 cancels into 16 two times, and 5 * 2 = 10.

EXAMPLE 4:

Solve for x if $- 7x = 42$ Number on right side of equation, goal 2 is achieved
                    Unknown on left side of equation but has a coefficient other than 1.

Take unknown's coefficient's multiplicative inverse.
The multiplicative inverse of - 7 is - 1/7.
Multiply both sides by multiplicative inverse
$(- 1/7)7x = 42(- 1/7)$
$x = - 6$ check solution in original algebraic equation
$(- 7)(- 6) = 42$
$42 = 42$ checked $x = - 6$ is the solution to the algebraic equation $- 7x = 42$. ■

Multiplying by - 1/7 is the same as dividing by - 7.

EXAMPLE 5:

Solve for x if $- x = - 26$ Number on right side of equation, goal 2 is achieved
                    Unknown on left side of equation but has a coefficient other than 1.

Take unknown's coefficient's multiplicative inverse.
The multiplicative inverse of - 1 is - 1.
Multiply both sides by multiplicative inverse
$- 1(- x) = - 26(- 1)$
$x = 26$ check solution in original algebraic equation
$-1(- 26) = 26$
$26 = 26$ checked $x = - 26$ is the solution to the algebraic equation $- x = 26$. ■

In reality, just change the signs on both sides of the equal sign whenever an algebraic equation results in - x equals some number.

Now let us end this section with combining like terms.

EXAMPLE 6:

Solve for x if $3x + 8x = 44$ Number on right side of equation, goal 2 is achieved
More than one unknown on left side of equation, combine
like terms

$11x = 44$ Take unknown's coefficient's multiplicative inverse.
The multiplicative inverse of 11 is 1/11.
Multiply both sides by multiplicative inverse
$(1/11)11x = 44(1/11)$
$x = 4$ check solution in original algebraic equation
$3 * 4 + 8 * 4 = 44$ multiply
$12 + 32 = 44$ add
$44 = 44$ checked $x = 4$ is the solution to the algebraic equation $3x + 8x = 44$. ■

Multiplying by 1/11 is the same as dividing by 11.

PROBLEMS:

Solve for the unknown.

1) $2a = 28$ 2) $4a = 32$

3) $7g = -49$ 4) $-2a = 44$

5) $3a = -15$ 6) $-5n = -20$

7) $-3a = -21$ 8) $a/2 = 6$

9) $b/7 = 9$ 10) $-s/8 = -4$

11) $(4/5)f = -16$ 12) $(-2/5)p = 22$

13) $7a + 2a = 18$ 14) $15c - 2c = -13$

15) $-a = 2$ 16) $-f = -5$

SECTION 3.3
SOLVING ALGEBRAIC EQUATIONS USING
THE ADDITIVE AND MULTIPLICATIVE INVERSES

The best way to solve algebraic equations in this section is to use the additive inverse first, and then the multiplicative inverse. Unwanted fractions may occur if the multiplicative inverse is used first. Many people do not like to work with fractions. The additive inverse must be used first to avoid fractions.

EXAMPLE 1:

Solve for x if $6x - 5 = 7$ additive inverse of -5 is $+5$
$6x - 5 = 7$
   $+5$ $+5$ combine down
$6x = 12$ multiplicative inverse of 6 is 1/6
$(1/6)6x = 12(1/6)$ multiply both sides
$x = 2$ check in original algebraic equation
$6 * 2 - 5 = 7$ multiply
$12 - 5 = 7$ subtract
$7 = 7$ checked
2 is the solution for the algebraic equation $6x - 5 = 7$. ▪

EXAMPLE 2:

Solve for x if $5 - 4x = 13$ additive inverse of 5 is -5
$5 - 4x = 13$
-5 -5 combine down
$-4x = 8$ multiplicative inverse of -4 is $-1/4$
$(-1/4)(-4x) = 8(-1/4)$ multiply both sides
$x = -2$ check in original algebraic equation
$5 - 4(-2) = 13$ multiply
$5 + 8 = 13$ add
$13 = 13$ checked
-2 is the solution for the algebraic equation $5 - 4x = 13$. ▪

EXAMPLE 3:

Solve for x if $9 - x = 10$ the additive inverse of 9 is -9
$9 - x = 10$
-9 -9
$-x = 1$ the multiplicative inverse of -1 is -1
$(-1)(-x) = 1(-1)$
$x = -1$ check in original algebraic equation
$9 - (-1) = 10$ multiply   remember invisible 1
$9 + 1 = 10$ add

10 = 10 checked

- 1 is the solution of the algebraic equation 9 - x = 10. ■

Now let us end this section with combining like terms, and using the distributive property first.

EXAMPLE 4:

Solve for x if $7x - 3 + 9x + 6 = 4x + 51$ combine like terms on left

$16x + 3 = 4x + 51$ the additive inverse of $4x$ is - $4x$, and the additive inverse of + 3 is - 3

$16x + 3 = 4x + 51$

- $4x$ - 3   - $4x$ - 3 combine down

$12x = 48$ the multiplicative inverse of 12 is 1/12

$(1/12)12x = 48(1/12)$ multiply

$x = 4$ check in the original algebraic equation

$7 * 4 - 3 + 9 * 4 + 6 = 4 * 4 + 51$ multiply

28 - 3 + 36 + 6 = 16 + 51 subtract on left   add on right

25 + 36 + 6 = 67 add

67 = 67 checked

4 is the solution for the algebraic equation $7x - 3 + 9x + 6 = 4x + 51$. ■

EXAMPLE 5:

Solve for x if $6(5x + 3) = 7(4x + 6)$ use distributive property on both sides

$30x + 18 = 32x + 42$ the additive inverse of $32x$ is - $32x$, and the additive inverse of 18 is - 18

$30x + 18 = 32x + 42$

- $32x$ - 18 - $32x$ - 18 combine down

- $2x = 24$ the multiplicative inverse of - 2 is - 1/2

$(- 1/2)(- 2x) = 24(- 1/2)$ multiply

$x = - 12$ check in original algebraic equation

$6(5 * 12 + 3) = 7(4 * 12 + 6)$ multiply inside parentheses first

$6(60 + 3) = 7(48 + 6)$ add inside parentheses first

$6 * 63 = 7 * 54$ multiply

378 = 378 checked

12 is the solution for the algebraic equation $6(5x + 3) = 7(4x + 6)$ ■

All checks, in all three sections, were done by hand. Many instructors do not allow calculators for this stage of Algebra. Calculators can be used for a check if someone is self studying, and does not have a lot of time to spare.

PROBLEMS:

Solve for the unknown.

1) $2a + 3 = 7$

2) $7f - 2 = 5$

3) $4c + 7 = 19$

4) $7n + 2 = 16$

5) $6 - 2a = 44$

6) $2 - 3o = -10$

7) $2n = 5n + 3$

8) $2c = 20 - 8c$

9) $6b + 2 = 7b + 19$

10) $8a - 4 = a - 39$

11) $4a + 2 = 2a - 10$

12) $3b - 3 = 5b + 5$

13) $2i + 4 = 4i - 6$

14) $2j - 5 + 5j = 6 + 4j + 4$

15) $2a + 2 - 6a = 4 + 7a - 20$

16) $7c - 6 + 3c + 27 = 12c - 11$

17) $4s + 2(2s - 9) = 5s - 6$

18) $2(4a - 3) - 2a = a + 19$

19) $8(2z + 1) = 4(7z + 7)$

20) $9(2z - 1) = 3(z + 2)$

SECTION 3.4
SOLVING LITERAL EQUATIONS

There are two forms of literal equations. One form is an algebraic equation with at least two unknowns. The other form is a formula without any given variables. Solving literal equations is nothing more than rearranging an algebraic equation, or a formula to suit one's needs.

In many cases, a given formula will have to be rearranged to solve a particular problem. Let us look at the distance formula. The distance formula is $D = RT$ where D is distance, R is the rate of speed, and T is time. If time and rate are given, the problem is a simple multiplication problem. If distance and either time or rate are given, the problem is a division problem. Example 1 will illustrate the changing of the distance formula.

EXAMPLE 1:

a) Solve for R if $D = RT$. R is multiplied by T. Divide both sides by T to isolate R.
$D/T = RT/T$
$D/T = R$
Distance divided by time is the rate of speed.

b) Solve for T if $D = RT$. T is multiplied by R. Divide both sides by R to isolate T.
$D/R = RT/R$
$D/R = T$
Distance divided by the rate of speed is time. ∎

Dividing by an unknown is the same as dividing by a numerical coefficient. None of the rules have changed from the previous three sections. The only difference is that we are not solving for a specific number. We are only rearranging a formula, or algebraic equation.

EXAMPLE 2:

Solve for m if $y = mx + b$. This algebraic equation is the point slope of a line, and will be explained in Chapter 4.

$y = mx + b$ The additive inverse of $+ b$ is $- b$.
$- b$ $- b$ combine down
$y - b = mx$. Since y and b are unlike terms, $y - b$ must be written out as shown.

$y - b = mx$ Divide both sides by x to isolate m. $y - b$ must be placed in parentheses because the entire left side is being divided.

$(y - b)/x = mx/x$
$(y - b)/x = m$. ∎

EXAMPLE 3:

Solve for y if $2x + 3y = 6$. This is an algebraic equation in 2 variables, and will be explained in Chapters 4 and 5.

$2x + 3y = 6$ The additive inverse of $2x$ is $- 2x$.
$- 2x \qquad - 2x$
$3y = 6 - 2x$ Since 6 and $2x$ are unlike terms, $6 - 2x$ must be written out as shown.

$3y = 6 - 2x$ Divide both sides by 3 to isolate y. $6 - 2x$ must be placed in parentheses because the entire right side is being divided.

$3y/3 = (6 - 2x)/3$
$y = (6 - 2x)/3$ ∎

EXAMPLE 4:

Solve for p if $b (p + q) = d$. Use distributive property on left
$bp + bq = d$ Take the additive inverse of bq since p is needed. The additive inverse of $+ bq$ is $- bq$.

$bp + bq = d$
$\quad - bq \quad - bq$
$bp = d - bq$ Since d and bq are unlike terms, $d - bq$ must be written out as shown.

$bp = d - bq$ Divide both sides by b to isolate p. $d - bq$ must be placed in parentheses because the entire right side is being divided.

$bp/b = (d - bq)/b$
$p = (d - bq)/b$ ∎

PROBLEMS:

Solve for the indicated variable.

1) $\quad Bh = V$ (for h)

2) $\quad LW = A$ (for W)

3) $\quad 2a + 3b = 7$ (for b)

4) $\quad 6b + 7c = 5$ (for b)

5) $\quad a (d + f) = q$ (for e)

6) $\quad b (h + i) = w$ (for i)

SECTION 3.5
WORD PROBLEMS WITHOUT FORMULAS

There are two types of word problems. The first type is a number problem that can be simply translated into an algebraic equation. The second type is a more complicated problem that can be translated with the aid of a chart. Only the first three examples in this section will be straight forward translations. The remaining examples and all the examples in the next section will be charted.

EXAMPLE 1:

The sum of twice an integer and 6 is 18. What is the integer?

STEP 1: Translate sentence into an algebraic equation.

$2x + 6 = 18$ Recall that twice means 2 times.

STEP 2: Solve the algebraic equation.

$2x + 6 = 18$ The additive inverse of + 6 is - 6.
  $- 6$ $- 6$ combine down
$2x = 12$ The multiplicative inverse of 2 is 1/2
$(1/2)2x = 12(1/2)$ multiply
$x = 6$ check in original algebraic equation

$2 * 6 + 6 = 18$ multiply
$12 + 6 = 18$ add
$18 = 18$ checked.
18 is the integer. ■

Multiplying by 1/2 is the same as dividing by 2.

The next two examples deal with consecutive integers. Consecutive integers are integers that follow one another. 1, 2, 3, etc. are consecutive positive integers. - 1, - 2, - 3, etc. are consecutive negative integers. - 1, 0, 1, etc. are consecutive integers. The first integer is always represented by x, or any other variable. $x, x + 1$ would be the representation of two consecutive integers. $x, x + 1, x + 2$ would be the representation of three consecutive integers. Just add 1 to get to the next consecutive integer.

EXAMPLE 2:

The sum of two consecutive integers is 43. What are the two integers?

STEP 1: Translate the sentence into an algebraic equation.
$x + (x + 1) = 43$. Parentheses are shown to show that two consecutive integers are being

added together.
STEP 2: Solve the algebraic equation.

$x + x + 1 = 43$ add like terms
$2x + 1 = 43$ The additive inverse of $+ 1$ is $- 1$.
   $- 1$ $- 1$ combine down
$2x = 42$ The multiplicative inverse of 2 is 1/2.
$(1/2)2x = 42(1/2)$ multiply
$x = 21$ check in original algebraic equation.

$21 + 21 + 1 = 43$ add
$43 = 43$ checked
21 is the first number. 22 must be the second number because 22 is next in line.

It is a good idea to double check solutions.
$21 + 22 = 43$ add
$43 = 43$ checked
21 and 22 are the consecutive integers. ■

        Consecutive even or odd integers are a little different than consecutive integers. $- 2, 0, 2, 4$, etc. are consecutive even integers. $- 3, - 1, 1, 3$, etc. are consecutive odd integers. The first integer is always represented by x, or any other variable. $x, x + 2$ would be the representation of two consecutive even or odd integers. $x, x + 2, x + 4$ would be the representation of three consecutive even or odd integers. Just add 2 to get to the next consecutive even or odd integer.

EXAMPLE 3:

The sum of two consecutive odd integers is 68. What are the two integers?

STEP 1: Translate the sentence into an algebraic equation.

$x + (x + 2) = 68$. Parentheses are shown to show that two consecutive odd integers are being added together.

STEP 2: Solve the algebraic equation.

$x + x + 2 = 68$ add like terms
$2x + 2 = 68$ The additive inverse of $+ 2$ is $- 2$.
   $- 2$ $- 2$ combine down
$2x = 66$ The multiplicative inverse of 2 is 1/2.
$(1/2)2x = 66(1/2)$ multiply
$x = 33$ check in original algebraic equation.

$33 + 33 + 2 = 68$ add

68 = 68 checked
33 is the first number. 35 must be the second number because 35 is next odd integer.
It is a good idea to double check solutions.
33 + 35 = 68 add
68 = 68 checked
33 and 35 are the consecutive odd integers. ∎

Some integer problems cannot be easily translated. These problems need a chart to help solve the problem. The chart method was taught to me by Professor Jane Sparks. The main idea of a chart is to organize the information from the word problem.

The chart consists of at least two columns, and two rows. The first column is for the type of information that one wants to find. The second column is for the algebraic representation of the information.

EXAMPLE 4:

One integer is 6 more than a second integer. If 4 times the smaller integer plus 5 times the larger integer is 102, find the integers.

STEP 1: find the information and chart it.

There is a small integer and a large integer that must be found.

Small integer
Large integer

STEP 2: Chart the algebraic representation for the information.
x must represent the small number because the other number is 6 more than.

Small integer x
Large integer $x + 6$

STEP 3: Reread problem to see if all information is charted.

4 times the smaller integer plus 5 times the larger integer equals 102. A third row and column must be added to the chart.

Small integer	x	$4x$	4 times the small
Large integer	$x + 6$	$5(x + 6)$	5 times the large
Total		102	Total goes in the column that is being added together.

STEP 4: Make an algebraic equation with the column that has the total in it.

$4x + 5(x + 6) = 102$

STEP 5: Solve the algebraic equation.

$4x + 5(x + 6) = 102$ Use the distributive property
$4x + 5x + 30 = 102$ Add like terms
$9x + 30 = 102$ The additive inverse of $+ 30$ is $- 30$
   $- 30$ $- 30$ combine down
$9x = 72$ The multiplicative inverse of 9 is 1/9
$(1/9)9x = 72(1/9)$ multiply
$x = 8$ x is the small integer. Now find the large integer by substituting in x.
$8 + 6 = 14$

Check to see if 4 times the small plus 5 times the large is 102.
$4 * 8 + 5 * 14 = 102$ multiply
$32 + 70 = 102$ add
$102 = 102$ checked

8 is the small integer, and 14 is the large integer. ■

Integer problems are not the only ones that can be charted.

EXAMPLE 5:

There were 56 more yes votes than no votes on an election measure. If 740 votes were cast in all, how many yes votes were there? How many no votes?

STEP 1: find the information and chart it.

The number of yes and no votes must be found.

Yes votes
No votes

STEP 2: Chart the algebraic representation for the information.
v represents the number of no votes because there are 56 more yes votes than no votes.

Yes votes $v + 56$
No votes v

STEP 3: Reread problem to see if all information is charted.

There were a total of 740 votes cast. A third row must be added to the chart.

Yes votes $v + 56$
No votes v
Total   740

STEP 4: Make an algebraic equation with the column that has the total in it.

$v + 56 + v = 740$

STEP 5: Solve the algebraic equation.

$v + 56 + v = 740$ Add like terms
$2v + 56 = 740$ The additive inverse of $+ 56$ is $- 56$
     $- 56$ $- 56$ combine down
$2v = 684$ The multiplicative inverse of 2 is 1/2
$(1/2)2v = 684(1/2)$ multiply
$v = 342$ v is the number of no votes. Now find the number of yes votes by substituting in v.

Check to see if the number of no votes and the number of yes votes total 740.
$342 + 398 = 740$ add
$740 = 740$ checked

342 is the number of no votes, and 398 is the number of yes votes. ■

EXAMPLE 6:

Barry worked twice as many hours as Carter. Trixie worked 11 more hours than Carter. If they worked a total of 35 hours, find out how many hours each worked.

STEP 1: find the information and chart it.

The number of Barry's, Carter's and Trixie's hours must be found.

Barry's hours
Carter's hours
Trixie's hours

STEP 2: Chart the algebraic representation for the information.

h represents the number of Carter's hours because there is no clue for the number of hours he worked. Barry worked twice as many hours as Carter. Trixie worked 11 hours more than Carter. Barry's and Trixie's hours can be translated into algebraic expressions.

Barry's hours $2h$ Twice Carter
Carter's hours h
Trixie's hours $h + 11$ 11 more than Carter

STEP 3: Reread problem to see if all information is charted.

There were a total of 35 hours worked. A third row must be added to the chart.

Barry's hours $2h$
Carter's hours h
Trixie's hours $h + 11$
Total hours   35

STEP 4: Make an algebraic equation with the column that has the total in it.

$2h + h + h + 11 = 35$

STEP 5: Solve the algebraic equation.

$2h + h + h + 11 = 35$ Add like terms
$4h + 11 = 35$ The additive inverse of $+ 11$ is $- 11$
    $- 11 - 11$ combine down
$4h = 24$ The multiplicative inverse of 4 is 1/4
$(1/4)4h = 24(1/4)$ multiply
$h = 6$ h is the number of hours Carter worked. Now find the number of hours Barry and Trixie worked by substituting in h.
Barry $2 * 6 = 12$
Trixie $6 + 11 = 17$

Check to see if the number of hours worked adds up to 35.
$6 + 12 + 17 = 35$ add
$35 = 35$ checked

Barry worked 12 hours, Trixie worked 17 hours, and Carter worked 6 hours. ■

EXAMPLE 7:

Jay has 5 more dimes than nickels. If the value of the coins is $4.10, how many dimes and nickels does he have?

STEP 1: find the information and chart it.

The number of dimes and nickels must be found.

Dimes
Nickels

STEP 2: Chart the algebraic representation for the information.
n represents the number of nickels because there are 5 more dimes than nickels.

Dimes	$n + 5$	5 more than nickels
Nickels	n	

STEP 3: Reread problem to see if all information is charted.

Total cash equals $4.10. A third row must be added to the chart.

Dimes	$n + 5$
Nickels	n
Total	$4.10

STEP 4: Make an algebraic equation with the column that has the total in it. The value of each coin must be multiplied to each expression. Each dime is worth ten cents. Each nickel is worth five cents.

$.10(n + 5) + .05n = 4.10$

STEP 5: Solve the algebraic equation.

$.10(n + 5) + .05n = 4.10$ Use distributive property
$.10n + .50 + .05n = 4.10$ Add like terms
$.15n + .50 = 4.10$ The additive inverse of $+ .50$ is $- .50$
$\quad\quad - .50 \quad\quad - .50$ combine down
$.15n = 3.60$ Multiply both sides by 100 to remove decimals.
$15n = 360$ The multiplicative inverse of 15 is 1/15
$(1/15)15n = 360(1/15)$ multiply
$n = 24$ n is the number nickels. Now find the number of dimes by substituting in n.
$24 + 5 = 29$ dimes

Check to see if the value of nickels and the value of dimes total $4.10.
.05 * 24 + .10 * 29 = 4.10 multiply
1.20 + 2.90 = 4.10 add
4.10 = 4.10 checked

24 is the number of nickels, and 29 is the number dimes. ■

An alternative procedure for Example 7 would be to divide both sides by .15 instead of multiplying both sides by 100 at $.15n = 3.60$. Either way the results would be the same. The reason why I decided to clear the decimals is because most people do not like working with decimals.

EXAMPLE 8:

Bill is 2 years less than 3 times as old as his sister. If the sum of their ages is 18 years, how old is Bill.

STEP 1: find the information and chart it.

Bill's age must be found
Bill's age

STEP 2: Chart the algebraic representation for the information.
a represents Bill's sister's age because Bill is 2 years less than 3 times his sister's age. Bill's sister's age must also be charted.

| Bill's age | $3a - 2$ | 2 years less than 3 times |
| Bill's sister's age | a | |

STEP 3: Reread problem to see if all information is charted.

Both ages total 18 years. A third row must be added to the chart.

Bill's age	$3a - 2$
Bill's sister's age	a
Total years	18

STEP 4: Make an algebraic equation with the column that has the total in it.

$3a - 2 + a = 18$

STEP 5: Solve the algebraic equation.

$3a - 2 + a = 18$ Add like terms
$4a - 2 = 18$ The additive inverse of -2 is $+2$
$\quad +2 \quad +2$ combine down
$4a = 20$ The multiplicative inverse of 4 is 1/4
$(1/4)4a = 20(1/4)$ multiply
$a = 5 \quad a$ is the age of Bill's sister. Now find Bill's age by substituting in a.
$3 * 5 - 2 = 13$

Check to see if the ages add up to 18
$13 + 5 = 18$ add
$18 = 18$ checked

Bill is 13 years old. ■

WARNING: Just because a was solved for does not mean that a is the answer to the problem. a represents the sister's age and not Bill's. The sister's age was needed to find Bill's. A common error is to stop once the unknown is found. Be careful. The first solution may not be the one that is wanted.

A polygon is a figure with at least three sides to it. Perimeters of any polygon can be found by simply adding up the sides if no formula is given.

EXAMPLE 9:

One side of a triangle is 5 meters longer than its base. The other side is 3 times the base. What are the lengths of the two sides and the length of the base of the triangle if the perimeter is 45 meters?

STEP 1: find the information and chart it.

The length of two sides, and the base must be found.

Length of side 1
Length of side 2
Length of the base

STEP 2: Chart the algebraic representation for the information.

L represents the length of the base because there is no clue for the base's length. Length 1 is 5 meters longer than the base. Length 2 is 3 times as long as the base. Length 1 and length 2 can be translated into algebraic expressions.

Length of side 1 $L + 5$ meters   5 meters longer than the base
Length of side 2 $3L$ 3 times as long as the base
Length of the base L

STEP 3: Reread problem to see if all the information is charted.

The perimeter is 45 meters. A fourth row must be added to the chart.

Length of side 1 $L + 5$ meters
Length of side 2 $3L$
Length of the base L
Perimeter   45 meters

STEP 4: Make an algebraic equation with the column that has the total in it.

$L + 5$ meters $+ 3L + L = 45$ meters

STEP 5: Solve the algebraic equation.

$L + 5$ meters $+ 3L + L = 45$ meters Add like terms
$5L + 5$ meters $= 45$ meters The additive inverse of $+ 5$ meters is $- 5$ meters
   $- 5$ meters $- 5$ meters combine down
$5L = 40$ meters The multiplicative inverse of 5 is 1/5
$(1/5)5L = 40$ meters $(1/5)$ multiply
$L = 8$ meters L is the length of the base. Now find the lengths of side 1 and side 2 by substituting in L.
Side 1 8 meters $+ 5$ meters $= 13$ meters
Side 2 $3 * 8$ meters $= 24$ meters

Check to see if the lengths of all three sides add up to 45.
8 meters $+ 13$ meters $+ 24$ meters $= 45$ meters add
45 meters $= 45$ meters checked

Side 1 is 13 meters long, side 2 is 24 meters long, and the base is 8 meters long. ■

     The word meters is not really needed to solve the problem. A meter is a unit of measure. Many instructors want the unit of measure included throughout the problem. The answer must be written as meters because meter is the unit of measure that is used. . Meters could be abbreviated as m.

The unit of measure must be included in the answer. A number is just a number without the unit of measure. The unit of measure tells exactly what the number represents.

PROBLEMS:

1) The sum of twice a number and 2 is 28. What is the number?

2) 4 times a number, minus 25, is 15. Find the number.

3) The sum of two consecutive integers is 17. Find the two integers.

4) The sum of three consecutive integers is 174. What are the three integers?

5) The sum of two consecutive even integers is 26. What are the two integers?

6) The sum of two consecutive odd integers is 44. What are the two integers?

7) One number is 3 more than another. If the sum of the smaller number and twice the larger number is 21, find the two numbers.

8) One number is 3 less than another. If 3 times the smaller number plus 5 times the larger number is 15, find the two numbers.

9) In an election, the winning candidate had 131 more votes than the loser. If the total number of votes cast were 1129, how many votes did each candidate receive?

10) A washer-dryer combination cost $346. If the washer cost $20 more than the dryer, what does each appliance cost?

11) Dilton has 2 more quarters than dimes. If the value of the coins is $2.60, how many dimes and quarters does he have?

12) Betty is 2 years less than twice as old as her sister Veronica. If the sum of their ages is 19 years, how old are both sisters?

13) Maria is 4 years older than Juan. The sum of their ages is 32 years. How old are both?

14) Jim is 8 years older than John. The sum of their ages is 46 years. How old are each now?

15) Betty earns $125 more per month than Frank. If their monthly salaries total $2730, what amount does each earn?

16) The Carter twins combined weight at birth was 12 pounds. If Frank weighed 2 more pounds than Martha, how much did each weigh?

17) On her vacation in South America, Ethel's expenses for food and lodging were $38 less than twice as much as her airfare. If she spent $1414 in all, what was her airfare?

18) One side of a triangle is 3 meters longer than its base. The other side is 7 times the base. What is the length of the base?

19) Two items are bought at a total price of $6.80. One of the items is $6.00 more than the other. How much does the less expensive item cost?

20) On an Algebra test, the highest grade was 42 points more than the lowest grade. The sum of the two grades was 138. Find the lowest grade.

21) Nancy runs a dairy farm. Last year her cow Bossie gave 375 gallons less than twice the amount produced by another cow, Bessie. The two cows gave 1464 gallons of milk. How many gallons of milk did Bessie give?

22) During a push up contest, Ralph did 35 more push ups than Clark did. The total number of push ups for both men was 177. Find the number of push ups that Clark did.

SECTION 3.6
WORD PROBLEMS WITH FORMULAS

Formulas aid us in solving problems. Units of measure are used with formulas. Treat units of measure as if they were variables. For instance (mi/h)h = mi because the h's cancel. A formula can tell how many rows or columns are needed in a chart. The formula for the perimeter of a rectangle ($2L + 2W = P$) will use three rows. The three rows will be length (L), width (W), and perimeter (P).

EXAMPLE 1:

The length of a rectangle is 2cm less than 4 times the width. If the perimeter is 56cm, find the dimensions of the rectangle.

STEP 1: find the information and chart it.

The length and the width must be found.

Length
Width

STEP 2: Chart the algebraic representation for the information.

W represents the width because the length is 2cm less than 4 times the width.

Length	$4W$ - 2cm	2cm less than 4 times the width
Width	W	

STEP 3: Reread problem to see if all information is charted.

The perimeter is 56cm. A third row must be added to the chart.

Length	$4W$ - 2cm
Width	W
Perimeter	56cm

STEP 4: Use the formula to make an algebraic equation with the column that has the perimeter in it.

$2L + 2W = P$
$2(4W$ - 2cm$) + 2W = 56$cm

$4W$ - 2cm must be enclosed in parentheses because $4W$ - 2cm represents the entire length.

STEP 5: Solve the algebraic equation.

$2(4W - 2\text{cm}) + 2W = 56\text{cm}$ Use the distributive property
$8W - 4\text{cm} + 2W = 56\text{cm}$ Add like terms
$10W - 4\text{cm} = 56\text{cm}$ The additive inverse of - 4cm is + 4cm
$\qquad + 4\text{cm} \quad + 4\text{cm}$ combine down
$10W = 60\text{cm}$ The multiplicative inverse of 10 is 1/10
$(1/10)10W = 60\text{cm} (1/10)$ multiply
$W = 6\text{cm}$ $\qquad W$ is the width. Now find the length substituting in W.
$L = 4(6\text{cm}) - 2\text{cm}$ multiply
$L = 24\text{cm} - 2\text{cm}$ subtract
$L = 22\text{cm}$

Check to see if the length and the width are correct by substituting in the formula.
$2(22\text{cm}) + 2(6\text{cm}) = 56\text{cm}$ multiply
$44\text{cm} + 12\text{cm} = 56\text{cm}$ add
$56\text{cm} = 56\text{cm}$ checked

The length of the rectangle is 22cm, and the width is 6cm. The abbreviation for centimeters is cm. ▄

      Another type of rectangle problem is when a sketch is given, and only one width, and one length is given. The procedure is still the same as Example 1.

EXAMPLE 2:

Find the dimensions of the given rectangle if the perimeter is 76 in.

$$3W + 2\text{in}$$

W

STEP 1: Chart the algebraic representation from the sketch.

Length $3W + 2\text{in}$
Width W

STEP 2: Reread problem to see if all information is charted.

The perimeter is 76in. A third row must be added to the chart.

Length $3W + 2$in
Width W
Perimeter   76in

STEP 3: Use the formula to make an algebraic equation with the column that has the perimeter in it.

$2L + 2W = P$
$2(3W + 2$in$) + 2W = 76$in

$3W + 2$in must be enclosed in parentheses because $3W + 2$in is the entire length.

STEP 4: Solve the algebraic equation.

$2(3W + 2$in$) + 2W = 76$in Use the distributive property
$6W + 4$in $+ 2W = 76$in Add like terms
$8W + 4$in $= 76$in The additive inverse of $+ 4$in is $- 4$in
      $- 4$in $- 4$in combine down
$8W = 72$in The multiplicative inverse of 8 is 1/8
$(1/8)8W = 72$in $(1/8)$ multiply
$W = 9$in W is the width. Now find the length substituting in W.
$L = 3(9$in$) + 2$in multiply
$L = 27$in $+ 2$in add
$L = 29$in

Check to see if the length and the width are correct by substituting in the formula.
$2(29$in$) + 2(9$in$) = 76$in multiply
58in $+ 18$in $= 76$in add
76in $= 76$in checked

The length of the rectangle is 22in, and the width is 6in. The abbreviation for inches is in.
■

Three columns are needed for the distance formula ($RT = D$). The distance formula uses columns instead of rows because the rows represent either people traveling along the same route, or one person going form one place to another and back again. It is helpful to make a drawing with any distance problem. The drawing helps to visualize the route being traveled.

EXAMPLE 3:

On Friday morning Joe drove from his house to the beach in 5hrs. In coming back on Sunday evening, heavy traffic slowed his speed by 8mi/h, and the trip took 6hrs. What was his average speed in each direction?

STEP 1: Make a drawing of the situation.

Joe's
House _____Beach

STEP 2: Enter the speed and time for going to the beach above the line, and the speed and time from the beach below the line.

Joe's R mi/h for 5hrs $\rightarrow$
House _____Beach

$\leftarrow R$ - 8mi/h for 6hrs

R represents the rate of speed.

STEP 3: create a chart for rate(R), and time (T)

	Rate	Time
Joe's house to beach	R	5hrs
Beach to Joe's house	R -8mi/h	6hrs

STEP 4: Use formula to add distance column.
$RT = D$

	Rate	Time	Distance
Joe's house to beach	R	5hrs	R (5hrs)
Beach to Joe's house	R -15mi/h	6hrs	$(R$ - 8mi/h)(6hrs)

STEP 5: Make an algebraic equation from the distance column. From the drawing, Joe traveled the same distance in both directions, therefore; the two distances equal each other.

R (5hrs) = (R - 8mi/h)(6hrs)

STEP 6: Solve for the algebraic equation

R (5hrs) = (R - 8mi/h)(6hrs) Use the distributive property.
R (5hrs) = R (6hrs) - 48mi The additive inverse of R (6hrs) is - R (6hrs).
- R (6hrs) - R(6hrs) combine down.
- R hr = - 48mi The multiplicative inverse of - hrs is - 1/hrs.
- R hr (- 1/hrs) = - 48mi (- 1/hr) multiply
R = 48mi/hr R is the rate of speed from Joe's house to the beach. Now find the rate of speed in the opposite direction by substituting in R.
48mi/h - 8mi/h = 40mi/hr.

To check this type of distance problem, use the formula for each rate of speed. The answers are correct if the distances are the same for each speed.

(48mi/h)5h = 240mi
(40mi/h)6h = 240mi

The distances are the same. Joe traveled 48mi/h to the beach, and traveled 40mi/h from the beach. ■

Now let us see how a distance problem is solved when there are two people traveling along the same route.

EXAMPLE 4:

Barbara leaves City A for City B at 11am driving at 45mi/h. At noon, Frank leaves city B for City A, driving at 50mi/h along the same route. If the cities are 330 miles are apart, what time will they pass each other?

STEP 1: Make a drawing of the situation.

City	Barbara →	City
A	_____	B

← Frank

STEP 2: Enter the speed and time for Barbara and Frank.

City	Barbara 45mi/h for T h →	City
A	_____	B

← Frank 50mi for T - 1h

T represents the time traveled. Frank has been on the road one less hour because he left an hour later.

STEP 3: create chart for rate(R), and time (T)

	Rate	Time
Barbara	45mi/h	T
Frank	50mi/h	T - 1h

STEP 4: Use formula to figure distance column.
$RT = D$

	Rate	Time	Distance
Barbara	45mi/h	T	(45mi/h)T
Frank	50mi/h	T - 1h	(50mi/h)(T - 1h)

STEP 5: Make an algebraic equation from the distance column. From the problem, the distance between city A and city B is 330mi. Where Barbara and Frank pass each other is somewhere within the 330mi. Barbara's plus Frank's distance make the entire distance.

(45mi/h)T + (50mi/h)(T - 1h) = 330mi

STEP 6: Solve for the algebraic equation

(45mi/h)T + (50mi/h)(T - 1h) = 330mi Use the distributive property.
(45mi/h)T + (50mi/h)T - 50mi = 330mi Add like terms
(95mi/h)T - 50mi = 330mi The additive inverse of - 50mi + 50mi.
       + 50mi   + 50mi combine down.
(95mi/h)T = 380mi The multiplicative inverse of 95mi/h is h/95mi.
(95mi/h)T (h/95mi) = (380mi)(h/95mi) multiply
T = 4h T is the time that Barbara has traveled. Now find the time that Frank has traveled by substituting for T.
4h - 1h = 3h

Barbara left at 11am and has been traveling for 4 hours. It should be 3pm by Barbara's watch. Frank left at noon and has been traveling for 3 hours. It should also be 3pm by Frank's watch. They pass each other at 3pm.

To check this type of distance problem, use the formula for each rate of speed and time. The answers are correct if the distances add up to the total distance.

Barbara (45mi/h)4h = 180mi
Frank (50mi/h)3h = 150mi
180mi + 150mi = 330mi. Add
330mi = 330mi. checked ▪

     There are other formulas to solve problems with, but the perimeter of a rectangle and the distance formula are the two most commonly used in a first algebra class.

PROBLEMS:

1) The length of a rectangle is 1 inch more than twice its width. If the perimeter of the rectangle is 98 inches, find the dimensions of the rectangle.

2) The length of a rectangle is 2 centimeters less than 3 times its width. If the perimeter of the rectangle is 28 centimeters, find the dimensions of the rectangle.

3) Find the dimensions of the given rectangle if the perimeter is 46 in.

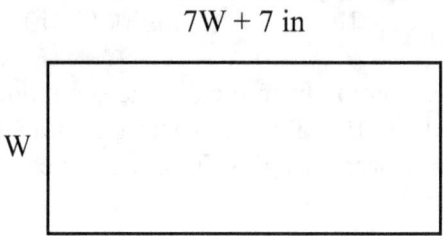

7W + 7 in

W

4) Find the dimensions of the given rectangle if the perimeter is 52 in.

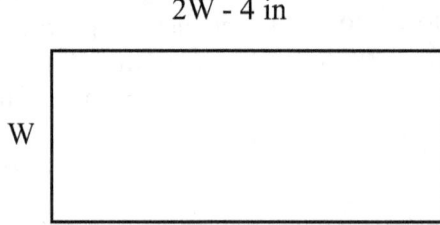

2W - 4 in

W

5) Archie drove 6 hours to attend a meeting. On the return trip, his speed was 9 miles per hour less, and the trip took 7 hours. What was his speed each way?

6) Diana rode her bike into the country for 2 hours. In returning, her speed was 5 miles per hour faster, and the trip took 1 hour. What was her speed each way?

7) At 6:00 AM, Dana leaves Yavin for Arkanis. She is traveling at a speed of 50mph. One hour later, Homer leaves Arkanis for Yavin on the same route, traveling at 59mph. When will Homer meet Dana if the two cities are 486 miles apart?

8) A train leaves Cerulean City for Vermilion City at 28mph. At the same time, another trains leaves Vermilion City for Cerulean City at 38mph. When will the two trains meet if the two cities are 264 miles apart?

SECTION 3.7
AN INTRODUCTION TO INEQUALITIES

An inequality usually means that two compared numbers are not equal to each other. 2 is not equal to 4 is an example of an inequality. An inequality can also mean something more than not equal to. Inequalities can also mean that a number is less than or greater than another number. 2 is less than 4 is also an example of an inequality.

There are two symbols used to show inequalities. $<$ is the symbol for less than. $>$ is the symbol for greater than. It is these two symbols that show inequality. The point of the inequality symbol always points to the smaller number. When reading an inequality, always read from left to right.

EXAMPLE 1:

a) $5 < 8$ 5 is less than 8   The inequality symbol points to the smaller number

b) $13 > 9$ 13 is greater than 9   The inequality symbol points to the smaller number
■

EXAMPLE 2:

Use $<$ or $>$ to compare the following pairs of numbers.

a)   6 _____ 24   6 is less than 24   The correct symbol is $<$

b)   65 _____ 12 65 is greater than 12   The correct symbol is $>$ ■

An algebraic inequality states that more than one number is less than or greater than a number. For example, $x < 10$ means that x can be any number that is less than 10. A graph can be used to show an algebraic inequality.

EXAMPLE 3:

Graph the algebraic inequality $x < 10$

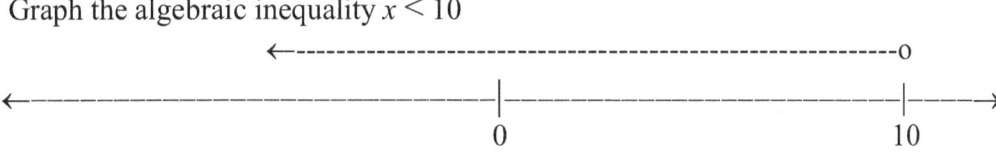

An open circle is drawn at 10 to indicate that 10 is not included. A line is then drawn to the left to indicate all the numbers are less than 10. The arrow at the end of the line indicates that there is an infinite number of numbers that are less than 10. ■

EXAMPLE 4:

Graph the algebraic inequality $x > -6$

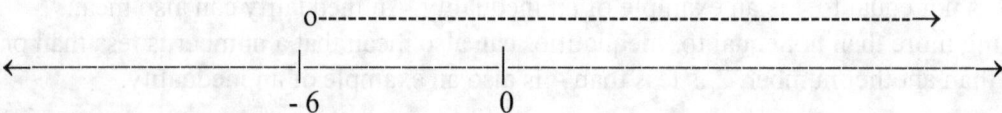

An open circle is drawn at - 6 to indicate that - 6 is not included. A line is then drawn to the right to indicate all the numbers are greater than - 6. The arrow at the end of the line indicates that there is an infinite number of numbers that are greater than - 6. ■

      Now let us expand our discussion of algebraic inequalities. An algebraic inequality can also mean that some number can be less than or equal to a number. x is less than or equal to 4 is an example of this type of inequality. The symbol for less than or equal to is $\leq$. Some number can also be greater than or equal to a number. x is greater than or equal to - 7 is an example of this type of inequality. The symbol for greater than is $\geq$. The given number is included with the addition of equal to in the inequality. These types of inequalities can also be graphed.

EXAMPLE 5:

Graph the algebraic inequality $x \leq 4$

A closed circle is drawn at 4 to indicate that 4 is included. A line is then drawn to the left to indicate all the numbers are less than 4. The arrow at the end of the line indicates that there is an infinite number of numbers that are less than 4. ■

EXAMPLE 6:

Graph the algebraic inequality $x \geq -7$

A closed circle is drawn at - 7 to indicate that - 7 is included. A line is then drawn to the right to indicate all the numbers are greater than - 7. The arrow at the end of the line indicates that there is an infinite number of numbers that are greater than - 7. ■

REMINDER: An open circle indicates that the given number is not included in the graph. A closed circle indicates that the given number is included in the graph.

The final example will show how to graph an inequality that lies between two integers.

EXAMPLE 7:

Graph the algebraic inequality $-7 \leq x \leq 4$

A closed circle is drawn at -7 and 4 to indicate that both -7 and 4 are included. A line is then drawn from -7 to 4 to indicate all the numbers between -7 and 4. ■

PROBLEMS:

Use < or > to compare the following pairs of numbers.

1)   2 _____ 27   2)   4 _____ -5

3)   0 _____ 7   4)   -2 _____ -3

Graph the following algebraic inequalities.

5) $x > 4$ 6) $x < 7$

7) $x > -7$ 8) $x \geq 2$

9) $x \leq -4$ 10) $x < 6$

11) $-2 \leq x \leq 5$ 12) $3 \leq x < 5$

SECTION 3.8
SOLVING ALGEBRAIC INEQUALITIES

Solving algebraic inequalities is almost identical to solving algebraic equations. The goals are the same. The unknown must be on the left of the inequality symbol, and numbers are on the right of the inequality symbol. The only differences are that division or multiplication by a negative number results in a change of direction for the inequality symbol, and the number that is found may or may not be used in the check. These differences will be explained in the appropriate examples.

EXAMPLE 1:

Solve for the algebraic inequality $x - 7 < 5$.

$x - 7 < 5$ The additive inverse of -7 is $+7$
$+7 \phantom{<}+7$ combine down
$x < 12$ check in original algebraic inequality

12 cannot be used in the check because 12 is not included in the inequality. Only numbers less than 12 can be used in the check. It does not matter which number is used as long as it is smaller than 12.

Let $x = 10$
$10 - 7 < 5$ subtract
$3 < 5$ check

Any number chosen that is less than 12 will give a true inequality. ■

EXAMPLE 2:

Solve for the algebraic inequality $5x > 7 + 4x$

$5x > 7 + 4x$ The additive inverse of $+4x$ is $-4x$
$-4x -4x$ combine down
$x > 7$ check in original algebraic inequality

7 cannot be used in the check because 7 is not included in the inequality. Only numbers less than 7 can be used in the check. It does not matter which number is used as long as it is larger than 7.

Let $x = 10$

$5 * 10 > 7 + 4 * 10$ multiply

$50 > 7 + 40$ add

$50 > 47$ check

Any number chosen that is greater than 7 will give a true inequality. ■

EXAMPLE 3:

Solve for the algebraic inequality $- 4x \geq 32$.

$- 4x \geq 32$ The multiplicative inverse of $- 4$ is $- 1/4$

$(- 1/4)(- 4x) \geq 32(- 1/4)$ multiply

$x \leq - 8$ check in original algebraic inequality
Notice the change in the inequality sign.

$- 8$ can be used in the check because $- 8$ is included in the inequality. It is also a good idea to check a number smaller than $- 8$ to see if it is indeed true inequality. Checking only $- 8$ only shows the equal part of the inequality.

Let $x = - 8$ Let $x = - 10$

$- 4(- 8) \geq 32$ multiply $- 4(- 10) \geq 32$ multiply

$32 \geq 32$ checked $40 \geq 32$ checked

Any number chosen that is less than $- 8$ will give a true inequality. ■

It was necessary to change the inequality sign because of the multiplication by a negative number. Let's reexamine $- 4x \geq 32$. Only negative numbers times a negative four will give positive numbers. Any number above $- 8$ will yield a false inequality. $- 4$ times $- 7$ is 28. 28 is not greater than or equal to 32. Only numbers smaller than $- 8$ could be used to make $- 4x \geq 32$ a true inequality.

Graphs can also be used in showing the results of an algebraic inequality.

EXAMPLE 4:

Solve the algebraic inequality $(3/4)x < 12$, and graph the solution.

$(3/4)x < 12$ The multiplicative inverse of $3/4$ is $4/3$

$(4/3)(3/4)x < 12(4/3)$ multiply

$x < 16$ check

Let $x = 8$ Use multiples of 4 to avoid fractions.
(3/4)8 < 12 multiply
6 < 12 checked.

Any number less than 16 will yield a true inequality.

Graph of the solution x < 16.

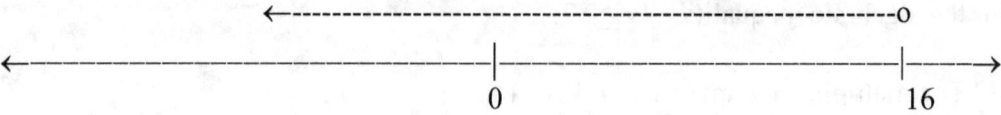

An open circle is drawn at 16 to indicate that 16 is not included. A line is then drawn to the left to indicate all the numbers are less than 16. The arrow at the end of the line indicates that there is an infinite number of numbers that are less than 16 that will make $(3/4)x$ < 12 a true inequality. ■

Any number can be used to check algebraic inequalities. 0, multiples of 5 and multiples of 10 are the easiest to use. Use multiples of the denominators if fractions are involved to avoid working with fractions.

PROBLEMS:

Solve for the following algebraic inequalities.

1) $x - 2 < 3$ 2) $x + 2 \geq 15$

3) $3x < 2x + 9$ 4) $2x - 5 \geq x + 2$

5) $3x + 2 < 2x - 3$ 6) $2x \leq 6$

7) $7x > -21$ 8) $-8x \geq 16$

9) $2x \geq 7x + 15$ 10) $3x + 5 > 9x - 13$

Solve and graph for the following algebraic inequalities.

11) $2x - 5 \geq x + 6$ 12) $4x + 4 < 3x - 2$

13) $2x \geq 6x + 12$ 14) $5x + 3 > 7x - 15$

CHAPTER 4
GRAPHING LINEAR ALGEBRAIC EQUATIONS
AND ALGEBRAIC INEQUALITIES IN 2 VARIABLES

SECTION 4.1
SOLUTIONS OF ALGEBRAIC EQUATIONS IN 2 VARIABLES

An algebraic equation in 2 variables is of the form $ax + by = c$. The coefficients of x and y, the variables, are represented by a and b respectively. c represents any real number. This section is divided into three parts. The first part deals with finding one variable when the other one is given. The second part deals with what are true solutions to given algebraic equations. The third part deals with finding solutions to algebraic equations.

Finding one variable when the other is given is just a matter of substitution, and solving an algebraic equation in one variable. Take the given variable, substitute it into the given algebraic equation, and solve for the remaining variable is all that is need to solve an algebraic equation in two variables.

EXAMPLE 1:

Solve for y when $x = 4$ for $x + y = 6$

$x + y = 6$ Substitute for x
$4 + y = 6$ The additive inverse of 4 is - 4
-4   - 4 combine down
$y = 2$ Check in original algebraic equation. Do not forget $x = 4$

$4 + 2 = 6$ Add
$6 = 6$ checked.

$y = 2$ is the missing variable. ■

The complete solution is written as (4,2). (4,2) is known as an ordered pair because x is the first term inside the parentheses and y is the second term. Writing solutions as ordered pairs is a convenience because when there are at least two variables, the work can become tedious, and the variables may be at different places in the work. The ordered pairs keep the variables together. Ordered pairs can also be used in a problem.

EXAMPLE 2:

Solve for the missing variable if (, 6) for $4x + y = 14$.

The first part of the ordered pair is missing. We must solve for x.

$4x + y = 14$ Substitute for y
$4x + 6 = 14$ The additive inverse of + 6 is - 6
   - 6   - 6 combine down

$4x = 8$ The multiplicative inverse of 4 is 1/4
$(1/4)4x = 8(1/4)$ multiply
$x = 2$ Check in original algebraic equation. Do not forget $y = 6$.

$4 * 2 + 6 = 14$ multiply
$8 + 6 = 14$ add
$14 = 14$ check

(2,6) is the solution to the algebraic equation $4x + y = 14$ ∎

      Finding which solutions are true in a given algebraic equation is just a simple substitution problem. If an ordered pair works then the ordered pair is a solution. If the ordered pair does not work then the ordered pair is not a solution.

EXAMPLE 3:

Which of the ordered pairs are solutions for $4x + 2y = 8$. (1,2), (5, - 7), (0,4)

$4x + 2y = 8$ Substitute an ordered pair into algebraic equation.
$4 * 1 + 2 * 2 = 8$ multiply
$4 + 4 = 8$ add
$8 = 8$ true.

$4x + 2y = 8$ Substitute another ordered pair.
$4 * 5 + 2(- 7) = 8$ multiply
$20 - 14 = 8$ subtract
$6 = 8$ false
$6 \neq 8$ ∴ (5, - 7) is not a solution

$4x + 2y = 8$ Substitute remaining pair.
$4 * 0 + 2 * 4 = 8$ multiply
$0 + 8 = 8$ add
$8 = 8$ true
 (1,2) and (0,4) are solutions for $4x + 2y = 8$ ∎

      Just pick any number for one variable and solve for the other one to solve algebraic equations when no variables are given. Pick numbers that are easy to work with.

EXAMPLE 4:

Find four solutions for $3x + 2y = 12$.

Let $x = 2$ Substitute into equation.
$3 * 2 + 2y = 12$ multiply
$6 + 2y = 12$ The additive inverse of 6 is - 6
$- 6$ - 6 combine down
$2y = 6$ The multiplicative inverse of 2 is 1/2
$(1/2)2y = 6(1/2)$ multiply
$y = 3$ Check in original algebraic equation. Do not forget $x = 2$
$3 * 2 + 2 * 3 = 12$ multiply
$6 + 6 = 12$ add
$12 = 12$ checked

Let $x = 4$ Substitute into equation.
$3 * 4 + 2y = 12$ multiply
$12 + 2y = 12$ The additive inverse of 12 is - 12
$- 12$ - 12 combine down
$2y = 0$ The multiplicative inverse of 2 is 1/2
$(1/2)2y = 0(1/2)$ multiply
$y = 0$ Check in original algebraic equation. Do not forget $x = 4$
$3 * 4 + 2 * 0 = 12$ multiply
$12 + 0 = 12$ add
$12 = 12$ checked

Let $x = - 6$ Substitute into equation.
$3(- 6) + 2y = 12$ multiply
$- 18 + 2y = 12$ The additive inverse of - 18 is + 18
$+ 18$ + 18 combine down
$2y = 30$ The multiplicative inverse of 2 is 1/2
$(1/2)2y = 30(1/2)$ multiply
$y = 15$ Check in original algebraic equation. Do not forget $x = - 6$
$3(- 6) + 2 * 15 = 12$ multiply
$- 18 + 30 = 12$ subtract
$12 = 12$ checked

Let $x = 0$ Substitute into equation.
$3 * 0 + 2y = 12$ multiply
$0 + 2y = 12$ Use additive identity
$2y = 12$ The multiplicative inverse of 2 is 1/2
$(1/2)2y = 12(1/2)$ multiply
$y = 6$ Check in original algebraic equation. Do not forget $x = 0$
$3 * 0 + 2 * 6 = 12$ multiply
$0 + 12 = 12$ add
$12 = 12$ checked

(2,3), (4,0), (- 6,15), and (0,6) are solutions to $3x + 2y = 12$. ◾

The secret to solving problems, such as; Example 4 is to pick numbers that will result in numbers on the right side of the equal sign that are divisible by the coefficient of the remaining variable. Notice that when the additive inverse was applied, the right side of the equal sign was divisible by 2.

PROBLEMS:

Solve for the missing variable.

1) $x + y = 10$, $x = 1$ 2) $x - y = 4$, $x = 10$

3) $6x + y = 3$, $y = -3$ 4) $x + 5y = 21$, $y = 0$

5) $2x - y = 12$, (, 0)   6) $x - 2y = 1$, (23,)

7) $4x - 5y = 15$, (5,)   8) $3x + 4y = 18$, (, 3)

Find the ordered pair that works for the following equations.

9) $6x - y = 6$. (2,6), (-3,-24), (0,1)   10) $x + 2y = 12$. (-12,0), (12,0)

11) $4x - y = 5$. (1,-1), (2,3), (0,5)   12) $x + 3y = 10$. (13,1), (13,-1)

13) $2x + y = 3$. (3,-3), (8,3), (0,1)   14) $x - 3y = 6$. (9,-1), (0,2)

15) $3x - 4y = 12$. (0,3), (4,0), (8,3)   16) $2x + 3y = 18$. (3,4), (12,-2)

Find four solutions for the following equations.

17) $3x - y = 6$ 18) $7x - y = 14$

19) $x + 9y = 18$ 20) $9x - 2y = 18$

SECTION 4.2
GRAPHING POINTS ON THE RECTANGULAR COORDINATE SYSTEM

Before points can be graphed on the rectangular coordinate system, one must understand the components of the rectangular coordinate system. The rectangular coordinate system is also known as the Cartesian coordinate system which is named after Rene' Descartes the French mathematician and philosopher who invented this system. This system is made up of four quadrants, two axes, and a center. The center is where the two axes intersect. The center is known as the origin. The axes are made up of one vertical axis, and one horizontal axis. The vertical axis is the y axis, and the horizontal axis is the x axis. The skeletal coordinate system is shown in Figure 4 - 1.

FIGURE 4 – 1

	y axis	
Quadrant 2	Quadrant 1	
	origin	
		x axis
Quadrant 3	Quadrant 4	

Each quadrant has a specific sign configuration. The sign configurations represent the ordered pairs that go into each quadrant. The origin has the ordered pair (0,0). If there is movement up from the origin then the y coordinate is positive. If there is movement down from the origin then the y coordinate is negative. If there is movement to the right of the origin then the x coordinate is positive. If there is movement to the left of the origin then the x coordinate is negative. The sign configurations, written as ordered pairs, are shown in Figure 4 - 2. The first sign is for the x coordinate, and the second sign is for the y coordinate.

FIGURE 4 - 2

	y axis	
Quadrant 2	Quadrant 1	
(-,+)	(+,+)	
		x axis
Quadrant 3	Quadrant 4	
(-,-)	(+,-)	

The placement of an ordered pair can be determined from Figure 4 - 2. Ordered pairs that have both positive *x* and *y* coordinates are placed in quadrant 1. Ordered pairs that have negative *x* and positive *y* coordinates are place in quadrant 2. Ordered pairs that have both negative *x* and *y* coordinates are placed in quadrant 3. Ordered pairs that have positive *x* and negative *y* coordinates are placed in quadrant 4.

To plot points, start at the origin and count the number of places in the *x* direction, and then move in the *y* direction. For example, to plot the point (4,5), start at (0,0) and move 4 spaces to the right, since *x* is positive, and then move 5 spaces up, since *y* is positive. Move in the opposite directions for negative coordinates.

EXAMPLE 1:

Plot the points A (4,5), B (3,-1), C (-4,-2), and D (-5,3)

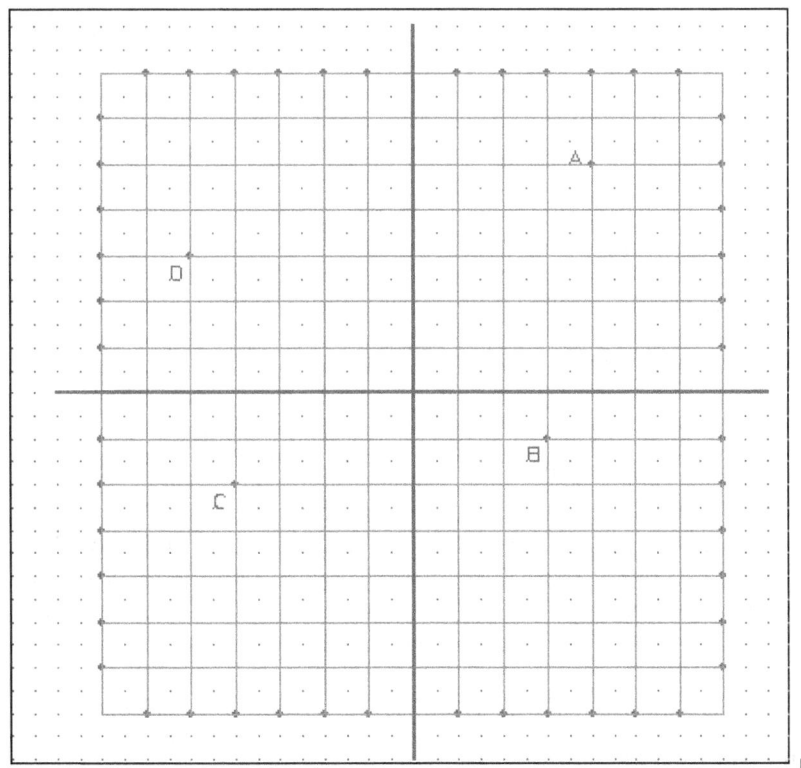

Suppose points are on a graph, and the ordered pairs were not given. Just count from the origin to the points to get the coordinates.

EXAMPLE 2:

What are the coordinates of the given points?

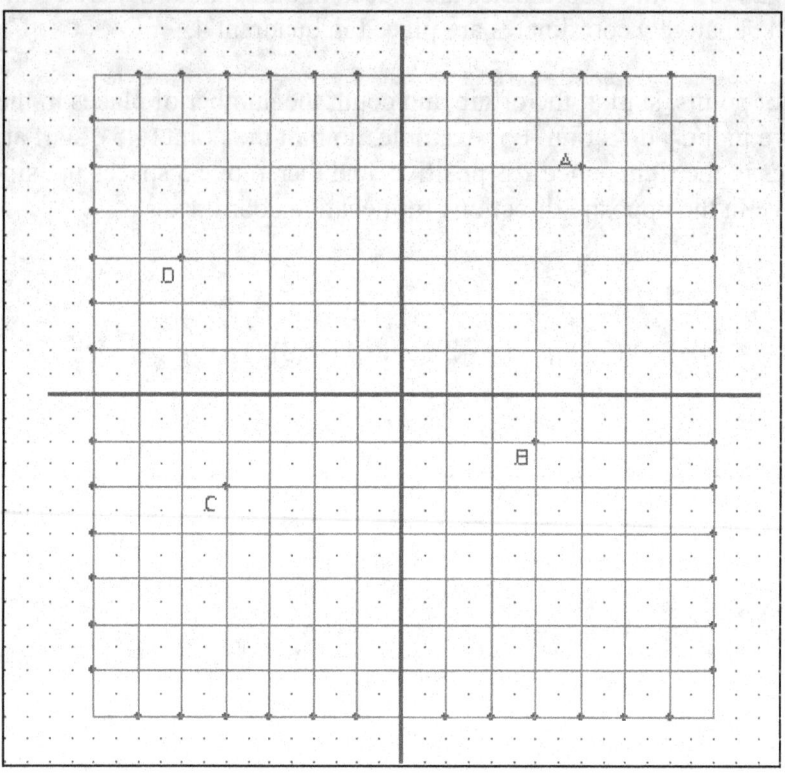

A is (4,5), B is (3,-1), C is (-4,-2), and D is (-5,3) ■

For every pair of points there is a midpoint. The midpoint is a point that is exactly between two points if a straight line was drawn between the two points. The formula for finding any midpoint is $M = ((x_1 + x_2)/2, (y_1 + y_2)/2)$. Be careful using the formula. If the coordinate signs are the same, add the coordinates together and divide by 2. If the coordinate signs are different, subtract the coordinates and divide by 2. Example 3 will show possible sign possibilities.

EXAMPLE 3:

a)   Find the midpoint of (3,1) and (11,1).
      The x coordinates are 3 and 11. Add
      $3 + 11 = 14$ Divide by 2
      $14/2 = 7$. The x coordinate of the midpoint is 7.

      The y coordinates are 1 and 1. Add
      $1 + 1 = 2$ Divide by 2
      $2/2 = 1$. The y coordinate of the midpoint is 1.

$M = (7,1)$

b)   Find the midpoint of (6,8) and (-2,-4)

The x coordinates are 6 and - 2. Subtract
6 - 2 = 4 Divide by 2
4/2 = 2. The x coordinate of the midpoint is 2.

The y coordinates are 8 and - 4. Subtract.
8 - 4 = 4 Divide by 2
4/2 = 2. The y coordinate of the midpoint is 2.

$M = (2,2)$

c)   Find the midpoint of (-4,-6) and (3,-3).
The x coordinates are - 4 and 3. Subtract
-4 + 3 = - 1 Divide by 2
- 1/2 = -1/2. The x coordinate of the midpoint is -1/2.
No division is needed because -1/2 is already in reduced form.

The y coordinates are - 6 and - 3. Add
- 6 - 3 = - 9 Divide by 2
- 9/2 = - 9/2. The y coordinate of the midpoint is - 9/2.
No division is needed because -9/2 is already in reduced form.

$M = (-1/2,-9/2)$

PROBLEMS:

The following points are found in what quadrant?

1) (9,3)   2)   (4, -9)

3) (-3,5)   4)   (-1,-2)

Plot the following points.

5) (1,2)   6)   (4, -1)

7) (-2,5)   8)   (-8,-6)

What are the coordinates of the given points?

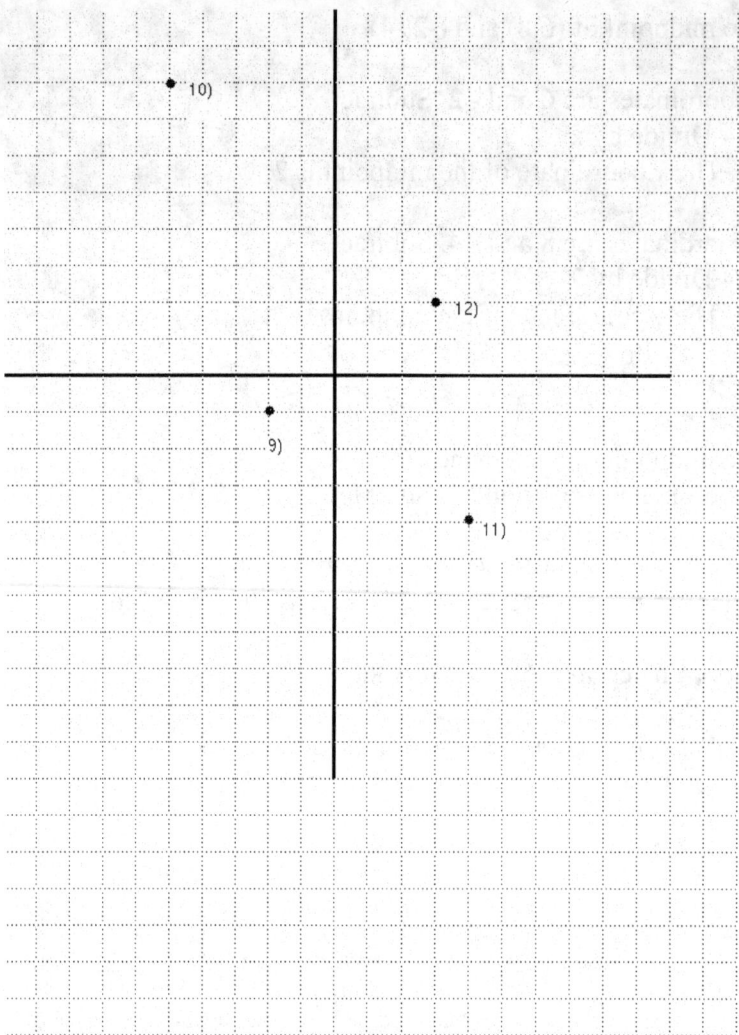

Find the following midpoints.

13)   (12, 19) and (1, 6)   14)   (10, 8) and (5, 0)

15)   (4, 10) and (3, 23)   16)   (7, 3) and (0, 7)

SECTION 4.3
GRAPHING LINEAR EQUATIONS IN 2 VARIABLES

All linear equations in two variables can be graphed as a straight line. The lines representing the equations can be horizontal, vertical, or diagonal. Diagonals are the most common while the horizontals and verticals are special cases where one of the variables remains fixed and the other one changes. Horizontal and vertical lines will be discussed at the end of this section.

Graphing a linear equation is just choosing a number for one of the variables, and solving for the other. Refer back to example 4 from section 4.1. At least two points are needed to draw a graph. A straight line can be drawn through the points. An arithmetic error has occurred if a straight line cannot be drawn though the points.

EXAMPLE 1:

Graph the line generated by the equation $x + y = 7$.

The easiest way to plot graphs with the equation in this form is to let $x = 0$, and $y = 0$.

Let $x = 0$. $x + y = 7$ becomes $0 + y = 7$. $y = 7$. One solution is (0,7).

Let $y = 0$. $x + y = 7$ becomes $x + 0 = 7$. $x = 7$. Another solution is (7,0).

Plotting the points (0,7), and (7.0) results in Graph 4.3a.

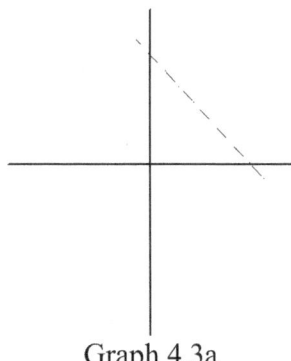

Graph 4.3a

Notice that the line goes through the two points. All straight lines have no end points. Pick a point along the line to check to see if this is a true representation of the equation. I choose to pick (3,4). $x + y = 7$ becomes $3 + 4 = 7$. $7 = 7$ is a true statement. The line is a true representation of $x + y = 7$. It is left up to the readers to pick other points along the line to check. ■

Now let us look at a linear equation that has a minus sign between the two variables.

102

EXAMPLE 2:

Graph the line generated by the equation $x - y = 5$.

The easiest way to plot graphs with the equation in this form is to let $x = 0$, and $y = 0$.

Let $x = 0$. $x - y = 5$ becomes $0 - y = 5$. $-y = 5$. $y = -5$. One solution is (0,-5).

Let $y = 0$. $x - y = 5$ becomes $x - 0 = 5$. $x = 5$. Another solution is (5,0).

Plotting the points (0,-5), and (5.0) results in Graph 4.3b.

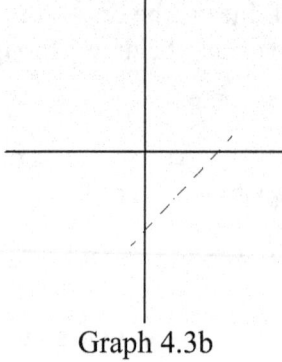

Graph 4.3b

Pick a point along the line to check to see if this is a true representation of the equation. I choose to pick (3,-2). $x - y = 5$ becomes $3 - (-2) = 5$. $3 + 2 = 5$. $5 = 5$ is a true statement. The line is a true representation of $x - y = 7$. It is left up to the readers to pick other points along the line to check. ∎

Recall that sign rules were used to solve the equation in example 2.

Now let us draw lines that have coefficients in the equations.

EXAMPLE 3:

Graph the line generated by the equation $3x + 2y = 12$.

Use $x = 0$, and $y = 0$ if both coefficients divide into the sum. Recall that the sum is the number that results from an addition. 12 is the sum when $3x + 2y$ are added together. 12 can be divided by both 3 and 2. Let $x = 0$. $3x + 2y = 12$ becomes $3 \cdot 0 + 2y = 12$. $2y = 12$. $y = 6$. (0,6) is one of the points. Let $y = 0$. $3x + 2y = 12$ becomes $3x + 2 \cdot 0 = 12$. $3x = 12$. $x = 4$. (4,0) is another point. Plotting the points (0,6), and (4.0) results in Graph 4.3c.

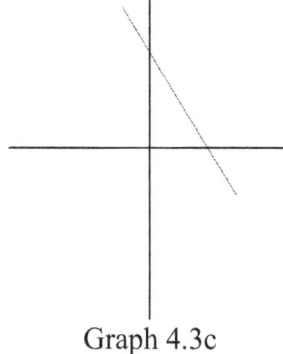

Graph 4.3c

Pick a point along the line to check to see if this is a true representation of the equation. I choose to pick (2,3). $3x + 2y = 12$ becomes $3 \bullet 2 + 2 \bullet 3 = 12$. $6 + 6 = 12$. $12 = 12$ is a true statement. The line is a true representation of $3x + 2y = 12$. It is left up to the readers to pick other points along the line to check. ∎

Now let us look at an equation where picking zero for the variables result in fractions.

EXAMPLE 4:

Graph the line generated by the equation $4x - 5y = 15$.

Let's try picking zero for the variables and see what happens. Let $x = 0$. $4x - 5y = 15$ becomes $4 \bullet 0 - 5y = 15$. $-5y = 15$. $y = -3$. $(0, -3)$ is a point on the line. Let $y = 0$. $4x - 5y = 15$ becomes $4x - 5 \bullet 0 = 15$. $4x = 15$. $x = 15/4$. $(15/4,0)$ is another point on the line.

Many students do not like working with fractions. Try finding a number that will result in a number on the right side of the equal sign that is divisible by the coefficient of the remaining variable. Let $x = 5$ since we started picking numbers for x. $4x - 5y = 15$ becomes $4 \bullet 5 - 5y = 15$. $20 - 5y = 15$. $-5y = -5$. $y = 1$. $(5, 1)$ is a point on the line. Plotting the points $(0,-3)$, and (5.1) results in Graph 4.3d.

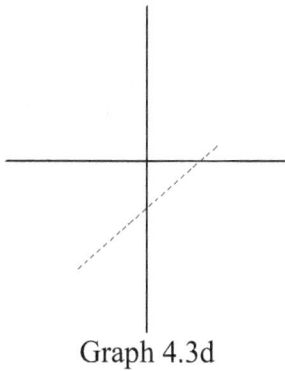

Graph 4.3d

Pick a point along the line to check to see if this is a true representation of the equation. I choose to pick (-5,-7). $4x - 5y = 15$ becomes $4(-5) - 5(-7) = 15$. $-20 + 35 = 15$. $15 = 15$ is a true statement. The line is a true representation of $4x - 5y = 15$. It is left up to the readers to pick other points along the line to check. ∎

Finding points that will generate graphs similar to example four is the same as finding solutions in the same manner that was shown in example 4 from section 4.1.

The final two examples will show equations that will generate a vertical and horizontal line.

EXAMPLE 5:

Graph the line generated by the equation $x = 8$.

Lines of this type can be simply drawn. Just draw a vertical line at $x = 8$. The resulting graph should look like graph 4.3e.

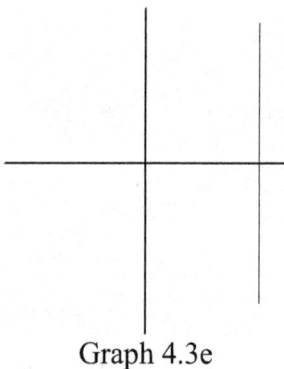

Graph 4.3e

A vertical line shows that x is fixed while y is changing. The points (8,3) and (8,-5) would be considered solutions to $x = 8$. ∎

EXAMPLE 6:

Graph the line generated by the equation $y = -6$.

Lines of this type can be simply drawn. Just draw a horizontal line at $y = -6$. The resulting graph should look like graph 4.3f.

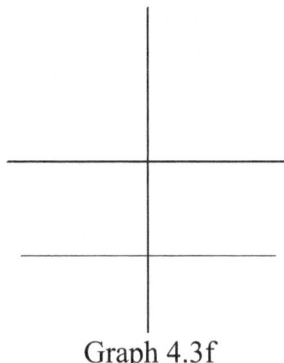

Graph 4.3f

A horizontal line shows that y is fixed while x is changing. The points $(8,-6)$ and $(-8,-6)$ would be considered solutions to $y = -6$.■

The vertical method for solving equations was not shown because students should be able to solve equations at this point in the book. Readers should refer back to chapter 3 for a reminder.

Here is a final note. The x axis is represented by the equation $y = 0$, and the y axis is represented by the equation $x = 0$. It is left up to the student to think about why these representations are true.

PROBLEMS:

Graph the lines generated by the following equations.

1) $x + y = 8$ 2) $x + y = 4$

3) $x - y = 8$ 4) $x - y = 4$

5) $4x + 2y = 8$ 6) $x + 5y = 5$

7) $4x - 2y = 8$ 8) $x - 5y = 5$

9) $8x - 5y = 17$ 10) $5x - 2y = 7$

11) $8x + 5y = 17$ 12) $5x + 2y = 8$

13) $x = -8$ 14) $x = 5$

15) $x = 4$ 16) $x = 1$

17) $y = -8$ 18) $y = -5$

19) $y = 6$ 20) $y = -2$

SECTION 4.4
GRAPHING LINEAR INEQUALITIES

All linear inequalities in two variables can be graphed as a straight line. The lines will either be solid or broken. Solid lines are drawn when inequalities also involve possible equalities. Broken lines are used for straight inequalities. $2x + 8y \leq 1$ would have a solid straight line. $2x + 8y > 1$ would have a broken line. The lines representing the inequalities can be horizontal, vertical, or diagonal. It seems that inequalities are the same as equations. Inequalities are basically the same as equations with respect to drawing the line. The main difference is in the solution.

More than one point can make an inequality a true statement. These points are on either side of the line. Points on the line make inequalities true statements when the inequality also includes equality. Points along the line that is formed by $2x + 8y \leq 1$ would make the inequality a true statement. Points along the line formed by $2x + 8y > 1$ would not make the inequality a true statement. Recall the open and closed circles from chapter 3. Broken and solid lines work in the same manner. The examples should clear up any confusion from the text.

EXAMPLE 1:

Graph the inequality generated by $x + y < 7$.

First pretend that the inequality is equality.

Use the techniques learned in the previous section.

Let $x = 0$. $x + y = 7$ becomes $0 + y = 7$. $y = 7$. One point on the line is (0,7).

Let $y = 0$. $x + y = 7$ becomes $x + 0 = 7$. $x = 7$. Another point on the line is (7,0).

Plotting the points (0,7), and (7.0) results in Graph 4.4a1 with a broken line.

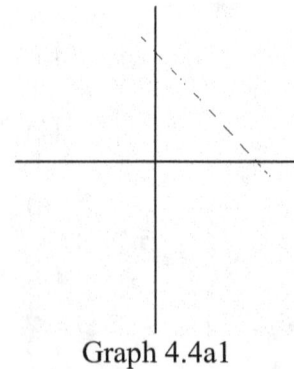

Graph 4.4a1

Now pick a point to see what side of the line needs to be shaded. I chose to pick (0,0).

The origin is the easiest to use. The side where the origin is gets shaded if the origin makes the inequality true. The other side of the line gets shaded if the origin does not make the inequality true. (0,0) makes $x + y < 7$ become $0 + 0 < 7$. $0 < 7$ is a true inequality. The area that includes (0,0) is shaded. The final graph should look like graph 4.4a2. It is left up to the readers to pick other points in the graph to check.

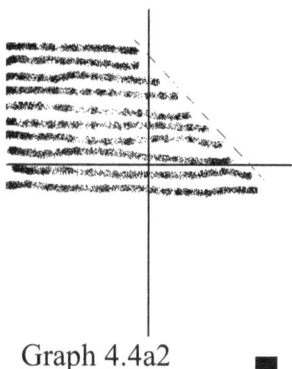

Graph 4.4a2

Now let us look at an inequality that has a minus sign between the two variables.

EXAMPLE 2:

Graph the inequality generated by $x - y > 5$.

First pretend that the inequality is equality.

Let $x = 0$. $x - y = 5$ becomes $0 - y = 5$. $y = 5$. One point on the line is (0,5).

Let $y = 0$. $x - y = 5$ becomes $x - 0 = 5$. $x = 5$. Another point on the line is (5,0).

Plotting the points (0,5), and (5.0) results in Graph 4.4b1 with a broken line.

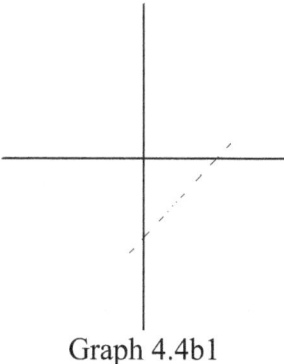

Graph 4.4b1

Now pick a point to see what side of the line needs to be shaded. I chose to pick (0,0). (0,0) makes $x - y > 5$ become $0 - 0 > 5$. $0 > 5$ is a false inequality. The area that does not include (0,0) is shaded. The final graph should look like graph 4.4b2. It is left up to the

108

readers to pick other points in the graph to check

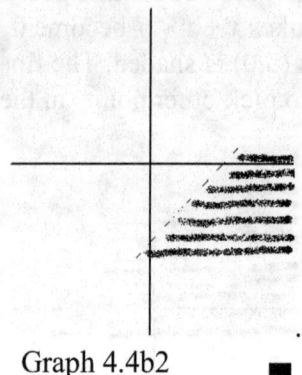

Graph 4.4b2

Now let us draw inequalities that have coefficients.

EXAMPLE 3:

Graph the inequality generated by $3x + 2y \leq 12$.

First pretend that the inequality is equality.

Let $x = 0$. $3x + 2y = 12$ becomes $3 \bullet 0 + 2y = 12$. $2y = 12$. $y = 6$. One point on the line is (0,6).

Let $y = 0$. $3x + 2y = 12$ becomes $3x + 2 \bullet 0 = 12$. $3x = 12$. $x = 4$. One point on the line is (4,0).

Plotting the points (0,6), and (4.0) results in Graph 4.4c1 with a solid line.

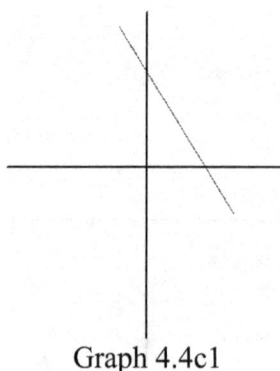

Graph 4.4c1

Now pick a point to see what side of the line needs to be shaded. I chose to pick (0,0). (0,0) makes $3x + 2y \leq 12$ become $3 \bullet 0 + 2 \bullet 0 \leq 12$. $0 + 0 \leq 12$. $0 \leq 12$ is a true inequality. The area that includes (0,0) is shaded. The final graph should look like graph 4.4c2. It is left up to the readers to pick other points in the graph to check

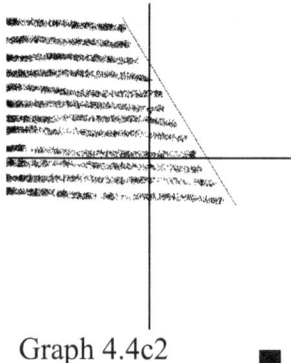

Graph 4.4c2

Now let us look at an equation where picking zero for the variables result in fractions.

EXAMPLE 4:

Graph the inequality generated by the equation $4x - 5y \geq 15$.

First pretend that the inequality is equality.

Let $x = 0$. $4x - 5y = 15$ becomes $4 \cdot 0 - 5y = 15$. $-5y = 15$. $y = -3$. $(0, -3)$ is a point on the line.

Let $y = 0$. $4x - 5y = 15$ becomes $4x - 5 \cdot 0 = 15$. $4x = 15$. $x = 15/4$. $(15/4, 0)$ is another point on the line.

Many students do not like working with fractions. Try finding a number that will result in a number on the right side of the equal sign that is divisible by the coefficient of the remaining variable. Let $x = 5$ since we started picking numbers for x. $4x - 5y = 15$ becomes $4 \cdot 5 - 5y = 15$. $20 - 5y = 15$. $-5y = -5$. $y = 1$. $(5, 1)$ is a point on the line. Plotting the points $(0, -3)$, and (5.1) results in Graph 4.3d with a solid line.

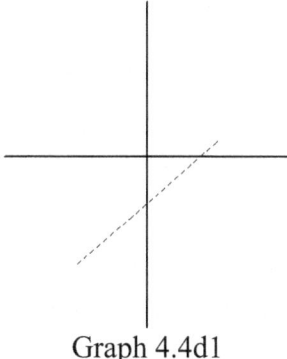

Graph 4.4d1

Now pick a point to see what side of the line needs to be shaded. I chose to pick $(0,0)$.

(0,0) makes $4x - 5y \geq 15$ becomes $4 \cdot 0 - 5 \cdot 0 \geq 15$. $0 - 0 \geq 15$. $0 \geq 15$ is a false inequality. The area that does not include (0,0) is shaded. The final graph should look like graph 4.4d2. It is left up to the readers to pick other points in the graph to check

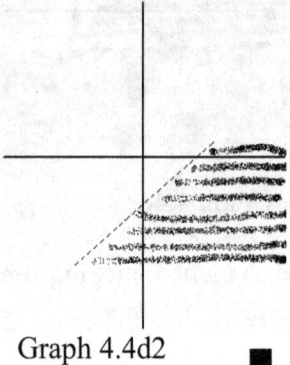

Graph 4.4d2

Finding points that will generate graphs similar to example four is the same as finding solutions in the same manner that was shown in example 4 from section 4.1.

The final two examples will show inequalities with a vertical and horizontal line.

EXAMPLE 5:

Graph the inequality generated by the $x < 8$.

Lines of this type can be simply drawn. Just draw a vertical broken line at $x = 8$. The resulting graph should look like graph 4.4e1.

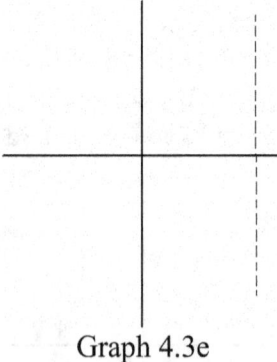

Graph 4.3e

Now pick a point to see what side of the line needs to be shaded. I chose to pick (0,0). (0,0) makes $x < 8$ becomes $0 < 8$. $0 < 8$ is a true inequality. The area that includes (0,0) is shaded. The final graph should look like graph 4.432. A y coordinate is not needed for the check since it is not part of the inequality. It is left up to the readers to pick other points in the graph to check

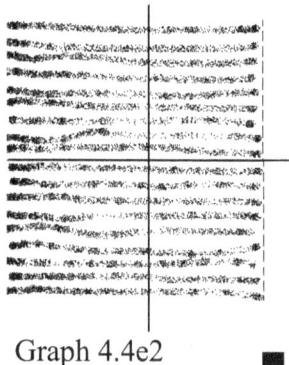

Graph 4.4e2

EXAMPLE 6:

Graph the line generated by the equation $y \geq -6$.

Lines of this type can be simply drawn. Just draw a horizontal solid line at $y = -6$. The resulting graph should look like graph 4.4f.

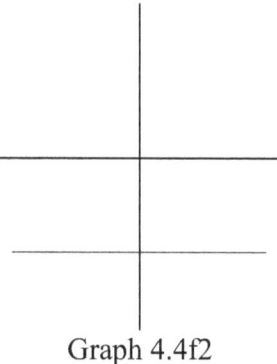

Graph 4.4f2

Now pick a point to see what side of the line needs to be shaded. I chose to pick (0,0). (0,0) makes $y \geq -6$ becomes $0 \geq -6$. $0 \geq -6$ is a true inequality. The area that includes (0,0) is shaded. The final graph should look like graph 4.4f2. An x coordinate is not needed for the check since it is not part of the inequality. It is left up to the readers to pick other points in the graph to check

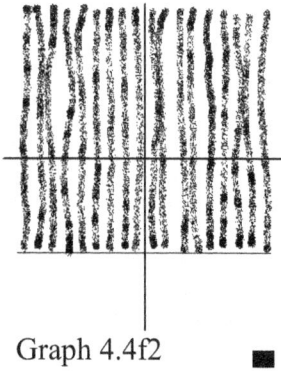

Graph 4.4f2

Here is a final note for those that are using this text as a supplemental text. A broken or horizontal line may go through the origin. Pick a point on either side of the line to see which side gets shaded.

PROBLEMS:

Graph the lines generated by the following equations.

1) $x + y < 8$ 2) $x + y > 4$

3) $x - y \le 8$ 4) $x - y \ge 4$

5) $4x + 2y < 8$ 6) $x + 5y > 5$

7) $4x - 2y \le 8$ 8) $x - 5y \ge 5$

9) $8x - 5y < 17$ 10) $5x - 2y > 7$

11) $8x + 5y \le 17$ 12) $5x + 2y \ge 8$

13) $x < -8$ 14) $x > 5$

15) $x \le 4$ 16) $x \ge 1$

17) $y < -8$ 18) $y > -5$

19) $y \le 6$ 20) $y \ge -2$

CHAPTER 5
SYSTEMS OF LINEAR EQUATIONS

SECTION 5.1
SOLVING SYSTEMS OF LINEAR EQUATIONS
GRAPHING METHOD

Drawing lines in systems of equations is exactly the same as drawing lines with an equation in two variables. Systems of equations have at least two equations and two variables. We will only deal with systems with exactly two equations and two variables. Systems with more equations and more variables will be covered in courses such as Intermediate Algebra, Linear Algebra, and Linear Systems. The graphing method shows solutions as a point of intersection, parallel lines, and a single line. The examples should clear up any confusion.

EXAMPLE 1:

Solve for the following system of equations using the graphing method: $x + y = 9$
$x - y = 3$

Graph the line generated by the equation $x + y = 9$.

Let $x = 0$. $x + y = 9$ becomes $0 + y = 9$. $y = 9$. One point on the line is $(0,9)$.

Let $y = 0$. $x + y = 9$ becomes $x + 0 = 9$. $x = 9$. Another point on the line is $(9,0)$.

Plotting the points $(0,9)$, and (9.0) results in Graph 5.1a

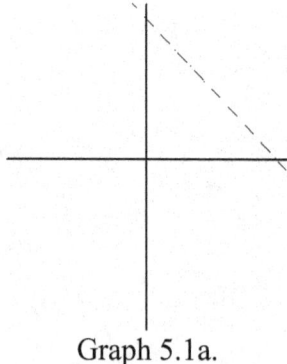

Graph 5.1a.

Now graph the line generated by the equation $x - y = 3$.

Let $x = 0$. $x - y = 3$ becomes $0 - y = 3$. $-y = 3$. $y = -3$. One point on the line is $(0,-3)$.

Let $y = 0$. $x - y = 3$ becomes $x - 0 = 3$. $x = 3$. Another point on the line is $(3,0)$.

Plotting the points $(0,-3)$, and (3.0) on graph 5.1a results in Graph 5.1b

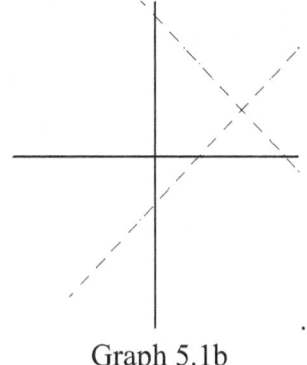

Graph 5.1b

The solution is the point of intersection which is the point (6,3). Check this in both of the original equations.

6 + 3 = 9 add   6 - 3 = 3 subtract
9 = 9 checked   3 = 3 checked

Both equations have lines that intersect at the point (6,3). ■

EXAMPLE 2:

Solve for the system of equations using the graphing method: $5x + 3y = 12$
                                                        $2x - 4y = 10$

Graph the line generated by the equation $5x + 3y = 12$.

Let $x = 0$. $5x + 3y = 12$ becomes $5 \cdot 0 + 3y = 12$. $3y = 12$. $y = 4$. One point on the line is (0,4).

Let $y = -1$. $5x + 3y = 12$ becomes $5x + 3(-1) = 12$. $5x - 3 = 12$. $5x = 15$. $x = 3$. Another point on the line is (3,-1).

Now graph the line generated by the equation $2x - 4y = 10$.

Let $x = 1$. $2x - 4y = 10$ becomes $2 \cdot 1 - 4y = 10$. $2 - 4y = 10$. $-4y = 8$. $y = -2$. One point on the line is (1,-2).

Let $y = 0$. $2x - 4y = 10$ becomes $2x - 4 \cdot 0 = 10$. $2x = 10$. $x = 5$. Another point on the line is (3,0).

Drawing the two lines result in Graph 5.2.

116

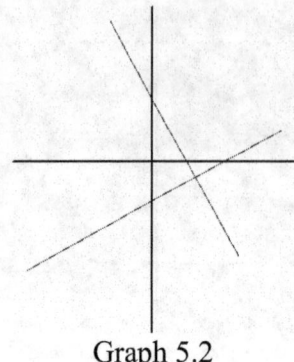

Graph 5.2

The solution is the point of intersection which is the point (3, - 1). Check in both equations.

5 * 3 + 3(- 1) = 12 multiply   2 * 3 - 4(- 1) = 10 multiply
15 - 3 = 12 subtract   6 + 4 = 10 add
12 = 12 checked   10 = 10 checked

Both equations have lines that intersect at the point (3, - 1).

EXAMPLE 3:

Solve for the system of equations using the graphing method: $4x + 6y = 14$
$6x + 9y = -6$

Graph the line generated by the equation $4x + 6y = 14$.

Let $x = -1$. $4x + 6y = 14$ becomes $4(-1) + 6y = 14$. $-4 + 6y = 14$. $6y = 18$. $y = 3$. One point on the line is $(-1,3)$.

Let $y = 1$. $4x + 6y = 14$ becomes $4x + 6(1) = 14$. $4x + 6 = 14$. $4x = 8$. $x = 2$. Another point on the line is $(2,1)$.

Now graph the line generated by the equation $6x + 9y = -6$.

Let $x = -1$. $6x + 9y = -6$ becomes $6(-1) + 9y = -6$. $-6 + 9y = -6$. $9y = 0$. $y = 0$. One point on the line is $(-1,0)$.

Let $y = 2$. $6x + 9y = -6$ becomes $6x + 9 \cdot 2 = -6$. $6x + 18 = -6$. $6x = -24$. $x = -4$. Another point on the line is $(-4,2)$.

Drawing the two lines result in Graph 5.3.

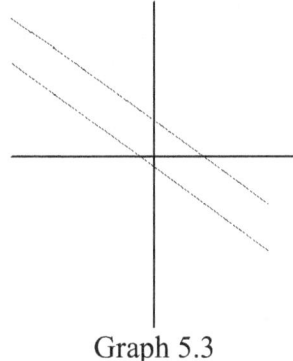

Graph 5.3

There is no solution. The lines are parallel. ▄

I will explain the numerical solution for parallel lines in the next section.

EXAMPLE 4:

Solve for the system of equations using the graphing method: $2x + 4y = -2$
     $4x + 8y = -4$

Graph the line generated by the equation $2x + 4y = -2$.

Let $x = 1$. $2x + 4y = -2$ becomes $2 \cdot 1 + 4y = -2$. $2 + 4y = -2$. $4y = -4$. $y = -1$. One point on the line is $(1,-1)$.

Let $y = 0$. $2x + 4y = -2$ becomes $2x + 4 \cdot 0 = -2$. $2x + 0 = -2$. $2x = -2$. $x = -1$. Another point on the line is $(-1,0)$.

Now graph the line generated by the equation $4x + 8y = -4$.

Let $x = 1$. $4x + 8y = -4$ becomes $4 \cdot 1 + 8y = -4$. $4 + 8y = -4$. $8y = -8$. $y = -1$. One point on the line is $(1,-1)$.

Let $y = 0$. $4x + 8y = -4$ becomes $4x + 8 \cdot 0 = -4$. $4x + 0 = -4$. $4x = -4$. x $= -1$. Another point on the line is $(-1,0)$.

Drawing the two lines result in Graph 5.4.

118

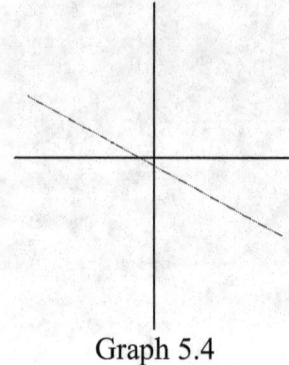

Graph 5.4

There are an infinite number of solutions. The lines intersect each other causing every point on the line to be a point of intersection. ■

I will explain the numerical solution for lines intersecting each other in the next section.

Now let us look at how a horizontal line intersects a vertical line.

EXAMPLE 5:

Solve for the system of equations using the graphing method: $x = 8$
$y = 1.$

Drawing the two lines result in graph 5.5.

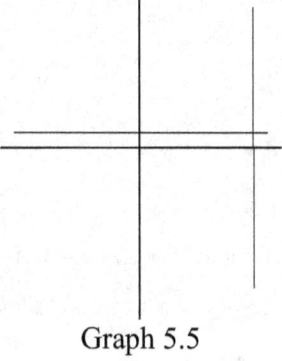

Graph 5.5

Both equations have lines that intersect at the point (8, 1). ■

Horizontal and vertical lines always intersect. The point of intersection is simply the ordered pair from the two equations. $x = 8$, and $y = 1$, from example 5, becomes the ordered pair (8,1) which is the point of intersection.

PROBLEMS:

Solve each of the following systems.

1) $x + y = 2$
    $x - y = 6$

2) $-x + y = 6$
    $x + y = 8$

3) $3x - y = 3$
    $-3x + 4y = 6$

4) $x + y = 7$
    $2x - 3y = 4$

5) $x - 2y = 1$
    $2x + y = 7$

6) $2x - 6y = 30$
    $6x + 7y = 15$

7) $3x + 4y = 4$
    $3x + 4y = 7$

8) $x + 2y = 13$
    $-2x - 4y = -26$

9) $-x + 4y = 2$
    $3x - 12y = 5$

10) $4x + 5y = 3$
    $2x - 5y = 9$

11) $2x - y = 2$
    $3x + y = 3$

12) $2x + y = 1$
    $4x - y = 3$

13) $x = 1$
    $y = 0$

14) $x = -2$
    $y = 7$

15) $x = -1$
    $y = -8$

16) $x = 2$
    $y = -1$

SECTION 5.2
SOLVING SYSTEMS OF LINEAR EQUATIONS
ELIMINATION METHOD

The elimination method is a more accurate method to solving systems of equations than the graphing method. In the graphing method, the point of intersection can only be exactly determined if the point of intersection falls on a point that has integers for coordinates. In the elimination method, exact points of intersections are found provided there is a point of intersection. The elimination method also gives the numerical results of inconsistent and dependant systems.

What is the elimination method? The elimination method is a method where one variable is eliminated, and the resulting equation is solved for the remaining variable. Once the remaining variable is found then that variable is substituted into one of the original two equations to find the other variable. Eliminating one variable can be as simple as combining the two equations.

EXAMPLE 1:

Solve for the following system of equations: $x + y = 9$
$$x - y = 3$$

The y's have opposite signs, and can be eliminated by combing the two equations.
$x + y = 9$
$x - y = 3$ combine down

$2x = 12$ Solve for x. The multiplicative inverse of 2 is 1/2.
$(1/2)x = 12(1/2)$ multiply
$x = 6$ Substitute into one of the original equations. It does not matter which one is chosen.

$6 + y = 9$ Solve for y. The additive inverse of 6 is - 6.
-6   - 6 combine down
$y = 3$

The solution is (6,3). Check this in both of the original equations.

$6 + 3 = 9$ add $6 - 3 = 3$ subtract
$9 = 9$ checked $3 = 3$ checked

Both equations have lines that intersect at the point (6,3). ∎

Equations can only be combined if one of the variables have opposite signs, and the coefficient is the same. In the above example, y had opposite signs, and a coefficient of 1. Let us now see what happens when there is opposite signs, but different coefficients.

EXAMPLE 2:

Solve for the system of equations: $5x + 3y = 12$
$2x - 4y = 10$

The y's have opposite signs, but have different coefficients. Since the equations cannot be combined to eliminate the y's, multiply each equation by some number that will make the y's coefficients the same. The easiest way to make the y coefficients the same is to multiply the top equation by the bottom coefficient, and multiply the bottom equation by the top coefficient. In other words, multiply the top equation by 4, and multiply the bottom equation by 3. I call this method of elimination a fail safe method. Solutions or non solutions always occur with the fail safe method of elimination.

$5x + 3y = 12$ multiply by 4
$2x - 4y = 10$ multiply by 3

$20x + 12y = 48$
$6x - 12y = 30$ combine down

$26x = 78$ The multiplicative inverse of 26 is 1/26
$(1/26)26x = 46(1/26)$ multiply
$x = 3$ Substitute into one of the original equations and solve for y.

$5* 3 + 3y = 12$ multiply
$15 + 3y = 12$ The additive inverse of 15 is - 15.
$- 15 - 15$ combine down
$3y = - 3$ The multiplicative inverse of 3 is 1/3
$(1/3)3y = - 3(1/3)$ multiply
$y = - 1$

The solution is (3, - 1). Check in both equations

$5 * 3 + 3(- 1) = 12$ multiply $2 * 3 - 4(- 1) = 10$ multiply
$15 - 3 = 12$ subtract $6 + 4 = 10$ add
$12 = 12$ checked $10 = 10$ checked

Both equations have lines that intersect at the point (3, - 1). ■

Now let us see what happens when there are no opposite signs present on the variables.

EXAMPLE 3:

Solve for the system of equations: $4x + 6y = 15$
$6x + 9y = -5$

There are no opposite signs on the variables. Since the equations cannot be combined to eliminate the y's or the x's, multiply each equation by some number that will make the y's or the x's coefficients the same. We can use the fail safe method to eliminate either y or x; however, one of the numbers must be a negative number to cause an elimination. Let's eliminate x this time. Multiply the top equation by 6, and multiply the bottom equation by - 4. It does not matter which number is made negative.

$4x + 6y = 15$ multiply by 6
$6x + 9y = - 5$ multiply by - 4

$24x + 36y = 90$
$- 24x - 36y = 20$ combine down

$0 = 180$ false statement ▄

The system of equations in example 3 resulted in a false statement. A false statement means that there is no point of intersection, and that there is no solution. Recall that <u>parallel lines have no point of intersection; therefore, this system of equations is an inconsistent system</u>. An inconsistent system is always the case when a result of 0 equals any real number occurs.

EXAMPLE 4:

Solve for the system of equations: $2x + 4y = - 2$
$4x + 8y = - 4$

Once again there are no opposite signs on the variables. Let us use the fail safe method to eliminate x. Multiply the top row by - 4, and the bottom row by 2.

$2x + 4y = - 2$ multiply by - 4
$4x + 8y = - 4$ multiply by 2

$- 8x - 16y = 8$
$8x + 16y = -8$ combine down

$0 = 0$ true statement. ▄

A result of 0 equals 0 means that every point on one line intersects every point on the other line. $0 = 0$ is a true statement. Any number that equals itself is considered a true statement. <u>Any system that winds up with a result of some number equaling itself would be considered a dependant system</u>. A dependent system has an infinite number of solutions because both lines are intersecting each other. A result of $5 = 5$ would also be considered a dependant system.

An alternative way to solve Example 4 is to multiply the top equation by - 2. The result would have been the same. I do not believe in showing short cuts at this level. A student may not be able to see a short cut during a test because of time constraints. I prefer giving a method that will work every time. Short cuts will work every time, but if a student can't see a short cut then there is another way for a student to solve system of equations every time.

It does not matter which variable is chosen for elimination to solve a system of equations. The key points to remember is that if the chosen variable has opposite signs, and the coefficients are the same then just combine the two equations. If the chosen variable has opposite signs then just use the fail safe method to change the equations. If the chosen variable has similar signs then use the fail safe method to change the equations, but remember to make one of the numbers a negative number.

PROBLEMS:

Solve each of the following systems.

1) $x + y = 2$
$x - y = 6$

2) $-x + y = 6$
$x + y = 8$

3) $3x - y = 3$
$-3x + 4y = 6$

4) $x + y = 7$
$2x - 3y = 4$

5) $6x + y = 42$
$2x + 3y = -2$

6) $2x - 6y = 30$
$6x + 7y = 15$

7) $3x + 4y = 4$
$3x + 4y = 7$

8) $x + 2y = 13$
$-2x - 4y = -26$

9) $-x + 4y = 2$
$3x - 12y = 5$

10) $4x + 5y = 14$
$6x + 7y = 10$

11) $2x + y = 8$
$4x + y = 9$

12) $2x + y = 1$
$4x - y = 3$

SECTION 5.3
SOLVING SYSTEMS OF LINEAR EQUATIONS
SUBSTITUTION METHOD

The substitution method is an alternate method to the elimination method for solving systems of equations. The results will come out the same. The substitution method is generally used when one of the two equations is set up like $y = 3x + 1$. $3x + 1$ would then be substituted into y in the other equation. Example 1 will clear up any confusion.

EXAMPLE 1:

Solve for the following system of equations: $y = -x + 9$
$$x - y = 3$$

y equals an expression in the first equation. $-x + 9$ can be substituted into y in the second equation.

$x - (-x + 9) = 3$ use distribution
$x + x - 9 = 3$ combine like terms
$2x - 9 = 3$ solve for x.   The additive inverse of -9 is +9

$2x - 9 = 3$
   $+ 9$ $+ 9$ combine down
$2x = 12$ The multiplicative inverse of 2 is 1/2.
$(1/2)x = 12(1/2)$ multiply
$x = 6$ Substitute into one of the original equations. It does not matter which one is chosen.

$6 - y = 3$ Solve for y. The additive inverse of 6 is - 6.
-6   - 6   combine down
$-y = -3$ The multiplicative inverse of -1 is -1
$-1(-y) = -1(-3)$ multiply
$y = 3$

The solution is (6,3). Check this in both of the original equations.

$3 = -6 + 9$ subtract $6 - 3 = 3$ subtract
$3 = 3$ checked $3 = 3$ checked

Both equations have lines that intersect at the point (6,3). ■

EXAMPLE 2:

Solve for the system of equations: $x + y = 10$
$$y = 4x$$

y equals an expression in the second equation. $4x$ can be substituted into y in the first equation.

$x + 4x = 10$ combine like terms

$5x = 10$ solve for x.   The multiplicative inverse of 5 is 1/5

$(1/5)5x = 10(1/5)$ multiply

$x = 2$ Substitute into one of the original equations. It does not matter which one is chosen.

$y = 4 \bullet 2$ Solve for y by multiplying.

$y = 8$

The solution is (2,8). Check this in both of the original equations.

$2 + 8 = 10$ add $8 = 4 \bullet 2$ multiply

$10 = 10$ checked $10 = 10$ checked

Both equations have lines that intersect at the point (2,8). ■

      Notice that I used the second equation to find y in example 2. The second equation was a simple multiplication. It is a good idea to use the easiest equation to find the second variable.

EXAMPLE 3:

Solve for the following system of equations: $y = 7x + 8$

$\qquad\qquad\qquad\qquad\qquad\qquad\qquad\quad x + 2y = 31$

y equals an expression in the first equation. $7x + 8$ can be substituted into y in the second equation.

$x + 2(7x + 8) = 31$ use distribution

$x + 14x + 16 = 31$ combine like terms

$15x + 16 = 31$ solve for x.   The additive inverse of +16 is -16

$15x + 16 = 31$

$\qquad$ - 16 -16 combine down

$15x = 15$ The multiplicative inverse of 15 is 1/15.

$(1/15)15x = 15(1/15)$ multiply

$x = 1$ Substitute into one of the original equations. It does not matter which one is chosen.

$y = 7 \bullet 1 + 8$ multiply

$y = 7 + 8$ add

$y = 15$

The solution is (1,15). Check this in both of the original equations.

$15 = 7 \bullet 1 + 8$	multiply	$1 + 2 \bullet 15 = 31$	multiply
$15 = 7 + 8$	add	$1 + 30 = 31$	add
$15 = 15$	checked	$31 = 31$	checked

Both equations have lines that intersect at the point (1,15). ■

The substitution method can be used for problems set up for the elimination method. To make an elimination problem into a substitution problem, the techniques for solving literal equations from chapter 3 are needed. Just solve for one of the variables, and then make a substitution into the other equation. I will not show how to turn an elimination setup into a substitution setup. Once a variable is solved, fractions can occur. My experience has been that most students want to avoid fractions. My advice to readers is do system of equations in the manner that the problems are set up. Do the elimination method if the problems are set up like section 5.2. Do the substitution method if the problems are set up like section 5.3. Some instructors specify which method to use on a test. Then there is no choice. Doing a method instead of the one an instructor specifies could result in not getting any credit for the problem even if the correct answer is obtained. I do not agree with this practice. In many cases, there are more than one way to solve a problem. Students should work out the best way to solve a problem. Just because the method is different does not mean the answer is wrong.

PROBLEMS:

Solve each of the following systems.

1) $y = -x + 2$
        $x - y = 6$

2) $-x + y = 6$
        $x = -y + 8$

3) $y = 3x - 3$
        $-3x + 4y = 6$

4) $y = -x + 7$
        $2x - 3y = 4$

5) $y = -6x + 42$
        $2x + 3y = -2$

6) $x = 3y + 15$
        $6x + 7y = 15$

7) $2x - 12y = 2$
        $x = 6y + 2$

8) $x = -2y + 13$
        $-2x - 4y = -26$

9) $x = 4y - 2$
        $3x - 12y = 5$

10) $x + y = 24$
        $y = 7x$

SECTION 5.4
SOLVING SYSTEMS OF LINEAR EQUATIONS
WORD PROBLEMS

The clue to solving word problems by systems of equations is that there are two equations and two totals. Setting up problems using the chart method will aid is solving word problems that deal with two equations and two totals.

EXAMPLE 1:

The sum of two integers is 10, and their difference is 2. What are the integers?

Here, the chart method is not necessary. Just translate into two equations.

$x + y = 10$
$x - y = 2$ combine down since the y's have opposite signs

$2x = 12$ solve for x.   The multiplicative of 2 is 1/2

$(1/2)2x = (1/2)12$ multiply
$x = 6$ substitute into one of the original equations.

$6 + y = 10$ solve for y.   The additive inverse of 6 is -6

  $6 + y = 10$
-6 -6 combine down
$y = 4$

The solution is (6,4). Check this in both of the original equations.

$6 + 4 = 10$ add $6 - 4 = 2$ subtract
$10 = 10$ checked $2 = 2$ checked

(6,4) are the only integers that work in both equations. ■

EXAMPLE 2:

Jay has $4.10 in dimes than nickels. If he has a total of 53 coins, how many dimes and nickels does he have?

STEP 1: find the information and chart it.

The number of dimes and nickels must be found.

Dimes
Nickels
STEP 2: Chart the algebraic representation for the information.
n represents the number of nickels, and d represents the number of dimes.

Dimes d
Nickels n

STEP 3: Reread problem to see if all information is charted.

Total cash equals $4.10, and total coins equals 53. A third row, and another column must be added to the chart. Remember, each dime is worth 10 cents, and each nickle is worth 5 cents.

	Coins	Total Value
Dimes	d	$.10d$
Nickels	n	$.05n$
Total	53	$4.10

STEP 4: Make an algebraic equation for both columns.

$$d + n = 53$$
$$.10d + .05n = \$4.10$$

STEP 5: Solve the system of equations.

First, clear out the decimals by multiplying the bottom row by 100.

$$d + n = 53$$
$$10d + 5n = 410$$

Second, multiply the top row by -10 so that the x will be eliminated.

$$-10d - 10n = -530$$
$$10d + 5n = 410 \qquad \text{combine down}$$

$-5n = -120$ solve for n.   The multiplicative inverse of -5 is -1/5

$(-1/5)(-5n) = -120(-1/5)$ multiply

$n = 24$ n is the number nickels. Now find the number of dimes by substituting in n.

$d + 24 = 53$ solve for d.   The additive inverse of $+24$ is -24.
$d + 24 = 53$
  $- 24$ $- 24$ combine down
$d = 29$

Check to see if the value of nickels, and the value of dimes total $4.10.
$.05 * 24 + .10 * 29 = 4.10$ multiply
$1.20 + 2.90 = 4.10$ add
$4.10 = 4.10$ checked

24 is the number of nickels, and 29 is the number dimes. ▪

EXAMPLE 3:

Angel bought 3 pens and 4 pencils for a total of $1.13. Buffy bought 9 pens and 10 pencils for a total of $2.23. What was the price per pen and pencil?

STEP 1: find the information and chart it.

The cost of a pen and pencil must be found.

pens
pencil

STEP 2: Chart the algebraic representation for the information.
x represents the number of pens, and y represents the number of pencils.

pens x
pencils n

STEP 3: Reread problem to see if all information is charted.

Angel and Buffy bought pens and pencils for two different total. Seperate colums for Angel and Buffy must be added to the chart along with a row for the two totals.

		Angel	Buffy
pens	x	$5x$	$9x$
pencils	y	$4y$	$10y$
Total		$1.13	$2.23

STEP 4: Make an algebraic equation for both columns.

$5x + 4y = 1.13$
$9x + 10y = 2.23$

STEP 5: Solve the system of equations.

Multiply the top row by 10, and the second row by -4. Remember the fail safe method from section 5.2.
$50x + 40y = 11.30$
$-36x - 40y = -8.92$ combine down
$14x = 2.38$ solve for x.   The multiplicative inverse of 14 is 1/14

$(1/14)(14x) = (1/14)2.38$ multiply

$x = 0.17$ x is the number of pens. Now find the number of pencils by substituting in x into either original equation.

$5 \bullet 0.17 + 4y = 1.13$ solve for y.   multiply
$0.85 + 4y = 1.13$ The additive inverse of 0.85 is -0.85

$0.85 + 4y = 1.13$
$-0.85 \quad - 0.85$ combine down
$4y = 0.28$ The multiplicative inverse of 4 is 1/4

$(1/4)4y = (1/4)0.28$ multiply
$y = 0.07$

A pen cost $0.17, and a pencil cost $0.07. The check is left to the reader. ■

EXAMPLE 4:

In a laboratory there are two chemical solutions that need to be combined to make a new solution. One solution is 10%, and the other is 30%. How much of each solution is needed to make a solution of 500 milliliters(ml) of 20%?

STEP 1: find the information and chart it.

The amount of each solution is needed.

10% solution
30% solution

STEP 2: Chart the algebraic representation for the information.
x represents the amount of 10% solution, and y represents the amount of 30 solution.

10% solution x
30% solution y

STEP 3: Reread problem to see if all information is charted.

A specific amount of each solution is needed to make the new solution. 10% of the first solution must be added to 30% of the second solution. Recall that that the word *of* means multiplication when *of* is involved with percentages. Both solutions must be set up as .10x and .30y. A second column and a new row is needed. The new row is for the totals.

10% solution x .10x
30% solution y .30y
Total   500ml .20 • 500ml

STEP 4: Make an algebraic equation for both columns.

$x +$ $y = 500$
.10x + .30y = .20 • 500   multiply on the right side of the equation to get a specific number in the second row.

$x +$ $y = 500$
.10x + .30y = 100

STEP 5: Solve the system of equations.

Multiply the top row by -.10 to eliminate the x.
-.10x - .10y = -50
.10x + .30y = 100   combine down

.20y = 100   solve for y.   The multiplicative inverse of .20 is 1/.20

(1/.20)(.20y) = (1/.20)20   multiply

$y = 20$ y represents the amount of 30% solution. Now find the amount of 10% solution by substituting in y into either original equation.

$x + 20 = 500$ solve for x.   The additive inverse of +20 is -20

$x + 20 = 500$
  -20   -20
$x = 480$

480ml of the 10% solution, and 20ml of the 30% solution is needed to make a solution of 20%. The check is left to the reader. ▄

I did not clear the decimals in examples 3 and 4. Many instructors allow calculators at this stage of Algebra. Anyone can still clear the decimals by multiplying the appropriate row by 100. Decimal problems are easier with a calculator.

EXAMPLE 5:

Diana won a lottery prize of $40,000. She wants to invest the entire amount into bonds with an interest rate of 12%, and into a savings account with an interest rate of 5%. What amount has she invested in each account if she receives interest of $2070 for one year?

STEP 1: find the information and chart it.

The amount of a each investment is needed.

Bonds   12%
Savings 5%

STEP 2: Chart the algebraic representation for the information.
b represents bonds, and s represents savings. Both are multiplied by their respective percentages because the amount that Diana invests into each one time the percentage is equal to the interest. Principle times rate equals interest. PR = I.

Bonds b 12% $.12b$
Savings s 5% $.05s$

STEP 3: Reread problem to see if all information is charted.

The total money invested and interest earned are known. These amounts are entered into the total row.

Bonds b 12% $.12b$
Savings s 5% $.05s$
Total $40,000 $2070

STEP 4: Make an algebraic equation for both columns.

$$b + \quad s = 40,000$$
$$.12b + .05s = 2070$$

STEP 5: Solve the system of equations.

multiply the top row by -0.05 to eliminate the y's

$-0.05b - 0.05s = -2000$
$.12b + .05s = 2070$ combine down

$.07b = 70$ solve for y.   The multiplicative inverse of .07 is 1/.07

$(1/.07)(.07b) = (1/.07)70$ multiply

$b = 1000$ b represents the amount of money invested in bonds. Now find the amount of money invested into a savings account by substituting in y into either original equation.

$1000 + s = 40,000$ solve for s.   The additive inverse of 1000 is -1000

$1000 + s = 40,000$
$-1000 \qquad -1000$ combine down
$s = 30,000$

     Diana invested $1000 in bonds, and $30,000 in a savings account. The check is left to the reader. ■

     I did not give checks in examples 3, 4, and 5 because readers should be able to check answers by themselves at this level.

PROBLEMS:

1)   The sum of two integers is 32, and their difference is 1. What are the two integers?

2)   Daisy has $3.00 in dimes and quarters. How many dimes and quarters does she have if she has 15 coins?

3)   Melvin has 27 coins for a total of $2.60. Melvin only has nickels and dimes. How many nickels and dimes does Melvin have?

4)   The cashier at the Abandoned Mine Gift Shop has some ten-dollar bills and some twenty-dollar bills. The total value of the money is $1480. If there is a total of 85 bills, how much of each type are there?

5)   A bank clerk has a total of 124 bills, both fives and tens. The total value of the money is $840, How many of each type of bill does the clerk have?

6) John bought 2 writing pads and 2 clipboards for a total of $4.72. Mary bought 3 writing pads and 4 clipboards for a total of $8.43. What was the price for one pad and one clipboard?

7) Kevin bought 2 highlighters and 6 markers for a total of $7.12. Jill bought 6 highlighters and 7 markers for a total of $9.81. What was the price for one highlighter and one marker?

8) Sally bought four candy bars and a pack of gum for a total of $2.75. Nancy bought two candy bars and three packs of gum for a total of $2.25. Find the cost of each.

9) Yugi bought five apples and four pears for a total of $4.81. Evan bought one of each of the same items. Find the cost of each.

10) A chemist has a 15%, and a 20% acid solution. How much of each solution should be used to make 130 milliliters(ml) of a 18% solution?

11) You have two alcohol solutions, one a 20% solution, and one a 35% solution. How much of each solution should be used to make 500ml of a 28% solution?

12) A pharmacist has a 15%, and a 30% alcohol solution. How much of each solution should be used to make 300 milliliters(ml) of a 25% solution?

13) Peter has a total of $10,000 invested in two accounts. One account pays 7% interest, and the other 5%. How much does he have in each account if his interest for one year is $670?

14) Vincent has $1000 to invest. Part of his money goes into bonds at 11% interest. The rest of his money goes into a savings account at 7%. How much money is in each of Vincent's accounts if he receives $615 in interest for 1 year?

15) Beth has $8000 to invest. Part of her money goes into bonds at 12% interest. The rest of her money goes into a savings account at 8%. How much money is in each of Beth's accounts if she receives $840 in interest for 1 year?

16) 700 tickets were sold for a concert. Ticket sales totaled $5000. Tickets were $6, and $8. How many of each ticket sold? **HINT:** Set this problem up like example 2.

17) Peanuts are selling for $1 a pound. Walnuts are selling for $2 a pound. How much of each type of nut is needed to make a mixture of 30 pounds that would sell for $1.75 a pound? **HINT:** Set this problem up like example 4.

CHAPTER 6
POLYNOMIALS

SECTION 6.1
INTRODUCING POLYNOMIALS

Polynomial is the formal name for an algebraic expression. Algebraic expressions were defined in chapter 1 as terms being separated by plus or minus signs. Polynomials are the same as algebraic expressions, but have special names when there are one to three terms involved. A single term polynomial is known as a monomial. A polynomial with two terms is known as a binomial. A polynomial with three terms is known as a trinomial. Word association can be used to help identify these three forms of polynomials. A monorail has one rail, a bicycle has two wheels, and a triangle has three sides are ways of remembering that a monomial has one term, binomials have two terms, and a trinomial has three terms. Polynomials with four or more terms do not have special names. Polynomials with four or more terms are considered ordinary polynomials.

EXAMPLE 1 (MONOMIALS):

a) x

b) $24yz$

c) 2 ■

Example 1 shows examples of monomials. Each one has a single term. A monomial can be any real number. Any real number is considered a single term. Variables need not be included in a monomial. Variables must be included when there are at least two terms present.

EXAMPLE 2 (BINOMIALS):

a) $x + y$

b) $2d + 3$

c) $x^2 - 9$ ■

EXAMPLE 3 (TRINOMIALS):

a) $x^2 + 6x + 9$

b) $2y^2 - 6y - 9$

c) $a^2 + 6ab + 9b^2$ ■

Polynomials can have more than one variable in them as was shown in examples 2 and 3.

EXAMPLE 4 (ORDINARY POLYNOMIALS):

a) $3x^3 + 4x^2 - 6x - 9$

b) $5y^8 + y^5 - 6y + 15$ ■

      It is important to discuss what types of algebraic expressions are not considered polynomials before discussing any further introductory topics. All polynomials are algebraic expressions. All algebraic expressions are not polynomials. The previous two statements may sound like a contradiction. There are certain terms that can not be considered as part of a polynomial. Any term that is a fraction, and has an unknown in the denominator cannot be considered a polynomial. The reasoning behind why fractions with unknowns in the denominator are not polynomials will be explained in chapter 9. It is my intention to only show the difference between polynomials and non-polynomials in this section.

      Algebraic expressions that have fractions with unknowns in the numerator are considered polynomials. This statement will also be explained in chapter 9.

EXAMPLE 5:

a) $\dfrac{x^2}{9}$ is a polynomial.

b) $\dfrac{1}{y}$ is not a polynomial.

c) $z^2 - \dfrac{z}{4} - 7$ is a polynomial.

d) $y^2 - \dfrac{2}{y} - 8$ is not a polynomial.

e) $\dfrac{x^2 + 9}{x^2 - 9}$ is not a polynomial. ■

      The next introductory topic deals with degrees. What is the degree of a polynomial? The degree of a polynomial can be easily determined when there is only one variable present. <u>The degree of a polynomial is simply the highest exponent present in the polynomial.</u> Degrees can be determined for polynomials that have more than one variable, but those discussions are more appropriate for courses that deal in polynomials with more than one variable. A first course in Algebra primarily deals with polynomials in one variable. There are cases when two variables are used, but these types of

138

polynomials will be discussed in chapter 8.

EXAMPLE 6:

a) $3x^3 + 4x^2 - 6x - 9$ is a third degree polynomial.

b) $5y^8 + y^5 - 6y + 15$ is an eighth degree polynomial.

c) $x^2 - 9$ is a second degree binomial or second degree polynomial.

d) y is a first degree monomial or first degree polynomial. ■

      Example 6d is a first degree polynomial because of the invisible 1 that is the exponent. Many students forget about the invisible 1, and claim that there is no degree. No degree polynomials will be discussed shortly. As long as there are variables with exponents, there will always be a degree.

      Non-degree polynomials are simply real numbers without variables. Any real number is considered a monomial. The degree for any real number is zero since there is no variable present.

EXAMPLE 7:

a)   4 has a degree of zero

b)   2.34 has a degree of zero

c)   - 235 has a degree of zero ■

      The last topic for this section is descending order. Descending order is simply a polynomial whose exponents go from high to low. $3x^3 + 4x^2 - 6x - 9$ is a polynomial in descending order. Descending order will play an important part with respect to long division (SECTION 6.4), and in factoring (CHAPTER 8).

      Polynomials that are not in descending order can be placed in descending order by using the commutative property. $5x^5 + x^8 - 6 + 15x$ is a polynomial that is not in descending order. $5x^5 + x^8 - 6 + 15x$ becomes $x^8 + 5x^5 + 15x - 6$ after the commutative property is applied. Notice that none of the signs were changed. The only change was that the polynomial is now in descending order.

PROBLEMS:

Label the following polynomials as monomial, binomial, trinomial, or ordinary

1) $3x$ 2) $4y + z$

3) $\quad 5y^2 + 5y + 6$

4) $\quad 6x^3 + 7x^2 - 8x - 9$

5) $\quad 10x^8 + x^2 - 3x + 4$

6) $\quad 9$

State whether the following are polynomials or not.

7) $\quad \dfrac{x^3}{10}$

8) $\quad \dfrac{1}{c}$

9) $\quad y^2 - \dfrac{3}{y}$

10) $\quad z^2 - \dfrac{z}{5}$

State the degree of the following polynomial.

11) $\quad 10x^3 + x^2 - 5x - 3$

12) $\quad 6y^7 + 9y^5 - 11y + 12$

13) $\quad y^2 - 13$

14) $\quad 77$

Put the following into descending order.

15) $\quad 10x^2 + x - 5x^3 - 3$

16) $\quad 6y + 9y^3 - 11y^4 + 12$

17) $\quad -5y - 2y^5 + 1 - y^2 + y^3$

18) $\quad 5x^5 + 2x^3 - 6x^4 + x^2 + 1 + 11x$

140

SECTION 6.2
ADDING AND SUBTRACTING POLYNOMIALS

Adding and subtracting polynomials without parentheses is the same as adding and subtracting like terms from chapter 1. It is not necessary to repeat what was already taught. Adding polynomials with parentheses is relatively easy. Just remove the parentheses, and combine like terms. Subtraction of polynomials with parentheses needs the use of the distribution property. A minus sign must be distributed through the parentheses. Example 1 will show addition, example 2 will show the distribution, and example 3 will show subtraction.

EXAMPLE 1:

a)   Add $3v - 7$ and $4v + 3$ rewrite
       $(3v - 7) + (4v + 3)$ the parentheses show that 2 binomials are being added, remove parentheses
       $3v - 7 + 4v + 3$ combine like terms using rules from chapter 2
       $7v - 4$ is the answer.

b)   Add $8t^2 - 5t + 4$ and $9t^2 + 9t - 1$ rewrite
       $(8t^2 - 5t + 4) + (9t^2 + 9t - 1)$ the parentheses show that 2 trinomials are being added, remove parentheses
       $8t^2 - 5t + 4 + 9t^2 + 9t - 1$ combine like terms using rules from chapter 2
       $17t^2 + 4t + 3$ is the answer

c)   Add $7x^3 - 3x^2 + x + 6$ and $2x^3 - 3x^2 - 5x + 6$ rewrite
       $(7x^3 - 3x^2 + x + 6) + (2x^3 - 3x^2 - 5x + 6)$ the parentheses show that 2 polynomials are being added, remove parentheses
       $7x^3 - 3x^2 + x + 6 + 2x^3 - 3x^2 - 5x + 6$ combine like terms using rules from chapter 2
       $9x^3 - 6x^2 - 4x + 12$ is the answer ■

REMINDER: There is an invisible 1 between the plus sign and the parenthesis which implies multiplication by 1.

The parentheses are removed in example 1 because a positive 1 multiplied to anything does not change any value.

EXAMPLE 2:

a) $- (5k + 4)$ distribute negative sign
       $- 5k - 4$ is the answer

b) $- (4z^2 + 7z - 6)$ distribute negative sign

        $- 4z^2 - 7z + 6$ is the answer

c) $- (-6r^3 + 5r^2 - 6r + 9)$

        $6r^3 - 5r^2 + 6r - 9$ is the answer ■

REMINDER: There is an invisible 1 between the negative sign and the parenthesis which implies multiplication by - 1.

When a negative 1 is multiplied to anything, all signs change according to the multiplication sign rules of chapter 2.

EXAMPLE 3:

a)   Subtract $3v - 7$ from $4v + 3$ rewrite, remember what comes after from is used first

        $(4v + 3) - (3v - 7)$ the parentheses show that 2 binomials are being subtracted, distribute negative sign

        $4v + 3 - 3v + 7$ combine like terms using rules from chapter 2

        $v + 10$ is the answer.

b)   Subtract $8t^2 - 5t + 4$ from $9t^2 + 9t - 1$ rewrite; remember what comes after from is used first

        $(9t^2 + 9t - 1) - (8t^2 - 5t + 4)$ the parentheses show that 2 trinomials are being subtracted, distribute negative sign

        $9t^2 + 9t - 1 - 8t^2 + 5t - 4$ combine like terms using rules from chapter 2

        $t^2 + 14t - 5$ is the answer

c)   Subtract $7x^3 - 3x^2 + x + 6$ from $2x^3 - 3x^2 - 5x + 6$ rewrite, remember what comes after from is used first

        $(2x^3 - 3x^2 - 5x + 6) - (7x^3 - 3x^2 + x + 6)$ the parentheses show that 2 polynomials are being subtracted, distribute negative sign

        $2x^3 - 3x^2 - 5x + 6 - 7x^3 + 3x^2 - x - 6$ combine like terms using rules from chapter 2

        $-5x^3 - 6x$ is the answer ■

In example 3c, notice that there is no x^2 term, nor a number at the end. Both terms went to zero. There is no need to write any term that has a zero in it.

Examples 1 and 3 all had like terms in them. Example 2 had unlike terms, but only a distribution was shown. Now let us add or subtract polynomials with unlike terms.

EXAMPLE 4:

a)   Add $y^2 + 7y$ and $y - 7$ rewrite
$(y^2 + 7y) + (y - 7)$ the parentheses show that 2 binomials are being added, remove parentheses
$y^2 + 7y + y - 7$ combine like terms using rules from chapter 2
$y^2 + 8y - 7$ is the answer.

b)   Add $3t^3 + 5t + 4$ and $5t^2 + 7t$ rewrite
$(3t^3 + 5t + 4) + (5t^2 + 7t)$ the parentheses show that a trinomial, and a binomial are being added, remove parentheses
$3t^3 + 5t + 4 + 5t^2 + 7t$ combine like terms using rules from chapter 2
$3t^3 + 5t^2 + 12t + 4$ is the answer

c)   Subtract $7x^3 + x + 6$ from $- 3x^2 - 5x + 6$ rewrite, remember what comes after from is used first
$(- 3x^2 - 5x + 6) - (7x^3 + x + 6)$ the parentheses show that 2 trinomials are being subtracted, distribute negative sign
$- 3x^2 - 5x + 6 - 7x^3 - x - 6$ combine like terms using rules from chapter 2
$-7x^3 - 3x^2 - 6x$ is the answer ■

REMINDER: When dealing with unlike terms, only add like terms, and just bring down the unlike terms without changing the signs.

PROBLEMS:

Add.

1) $2f - 6$ and $7f + 2$ 2) $3c^2 - 22c$ and $7c^2 - 2c$

3) $4h^2 - 6h$ and $-7h^2 + 4h$ 4) $3d^2 + 4d - 1$ and $6d^2 - 8d + 1$

Use distribution.

5) $- (3d + 5g)$ 6) $- (6f - 7h)$

7) $- (-7r^2 + 8r - 9)$ 8) $- (-3a^6 - 2a^4 + 8a^2)$

Subtract.

9) $k + 1$ from $3k - 4$ 10) $9j^2 - 3j$ from $5j^2 - 3j$

11) $7n^2 + 2n$ and $2n^2 + 3n$ 12) $m^2 - 3m - 4$ from $5m^2 - 6m - 7$

Add.

13) $4g^2 + 1$ and $2g + 2$

14) $2a^3 - 2a^2$ and $7a^2 - 4a$

15) $2b - 3$ and $- 4b^2 + 2$

16) $3d^2 + 4d - 8$ and $-2d^2 + 1$

Subtract.

17) $3c + 7$ from $8c^2 - 9c$

18) $3y^3 + y^2$ from $4y^3 - 5y$

19) $- 3b^2 + 4b - 7$ from $-8b^2 + b - 9$

20) $- 2h^4 - 3h^3 + 5h^2$ from $-9h^5 + 2h^4 - 3h^2 + 2h$

Do the indicated operations, and use order of operations if possible.

21) $(3g^2 - 7g + 2) - (10g^2 + 3g + 6)$

22) $(-8h^2 - 2h + 5) - (-8h^2 + h + 1)$

23) $((9z^2 - 3z + 5) - (3z^2 + 2z - 1)) - (z^2 - 2z - 3)$

24) $((5a^2 + 2a - 3) - (-2a^2 + a - 2)) - (2a^2 + 3a - 5)$

25) $((13d^2 + 7d - 9) - (7d^2 - 6d - 5)) - (5d^2 + 6d + 7)$

26) $((9k^2 - 6k + 7) - (6k^2 - 5k + 6)) - (-6k^2 - 7k + 9)$

27) $((17h^2 - 3h + 13) - (-11h^2 + 2h + 11)) - (-9h^2 - 2h - 11)$

28) $((j^2 - 10j + 2) - (8j^2 - 3j + 4)) - (-7j^2 - 5j + 12)$

144

SECTION 6.3
MULTIPLYING POLYNOMIALS

The same multiplication rules from chapter 2 are used when multiplying polynomials. The multiplication rule for multiplying monomials together is exactly the same as rule 1 of exponents. Multiplying a polynomial by a monomial is the distribution property. Multiplication with binomials and above will be the emphasis of this section.

Three types of results can occur when two binomials are multiplied together. One result is a trinomial which is the most common result. A binomial can occur after two binomials are multiplied together provided certain conditions are met. A four term polynomial can occur when there are no identical terms in either binomial being multiplied. All three results will be explained in their appropriate examples. One common procedure is used to get any of the three results. The procedure is called the foil method. Foil is an anagram for first, outside, inside, and last. Each letter represents what terms, from the two binomials, are multiplied together. F represents the first term in each binomial. O represents the outside terms. I represents the inside terms. L represents the last terms. (see figure 1) Each term gets multiplied together according to the letter that the terms fall on. After all four multiplications are completed; addition or subtraction of like terms takes place if possible.

REMINDER: Parentheses next to each other implies multiplication.

FIGURE 1

$(a + b)(c + d)$

EXAMPLE 1:

a) $(m + 5)(m + 8)$ multiply the terms on F: $mm = m^2$
multiply the terms on O: $8m$
multiply the terms on I: $5m$
multiply the terms on L: $8*5 = 40$
write all the terms as a polynomial and then add or subtract any like terms

$m^2 + 8m + 5m + 40 = m^2 + 13m + 40$

b) $(o - 9) (o - 3)$ multiply the terms on F: $oo = o^2$

                                          multiply the terms on O: $- 3o$

                                          multiply the terms on I: $- 9o$

                                          multiply the terms on L: $(- 3) (- 9) = 27$

                                          write all the terms as a polynomial and then add or subtract any like terms

$$o^2 - 3o - 9o + 27 = o^2 - 12o + 27$$

c) $(u + 2) (u - 9)$ multiply the terms on F: $uu = u^2$

                                          multiply the terms on O: $- 9u$

                                          multiply the terms on I: $2u$

                                          multiply the terms on L: $2*(- 9) = - 18$

                                          write all the terms as a polynomial and then add or subtract any like terms

$$u^2 - 9u + 2u - 18 = u^2 - 7u - 18$$

d) $(d - 9) (d + 8)$ multiply the terms on F: $dd = d^2$

                                          multiply the terms on O: $8d$

                                          multiply the terms on I: $- 9d$

                                          multiply the term on L: $- 9 * 8 = - 45$

                                          write all the terms as a polynomial and then add or subtract any like terms

$$d^2 + 8d - 9d - 45 = d^2 - d - 45 \ \blacksquare$$

     Two binomials multiplied together commonly result in a trinomial. The F multiplication becomes a square, the OI multiplication is combined, and the L multiplication is a product of two numbers. Trinomials will result even if the last terms are decimals, fractions, or if the last term is a different letter. It is left up to the reader to do the problems with the fractions, decimals, and different unknowns in the last term. The next two examples will show results that are not trinomials.

EXAMPLE 2:

a) $(v + 5) (v - 5)$ multiply the terms on F: $vv = v^2$

                                          multiply the terms on O: $- 5v$

                                          multiply the terms on I: $5v$

                                          multiply the terms on L: $5*(- 5) = - 25$

                                          write all the terms as a polynomial and then add or subtract any like terms

$$v^2 - 5v + 5v - 25 = v^2 - 25$$

b) $(g - r)(g + r)$ multiply the terms on F: $gg = g^2$
multiply the terms on O: gr
multiply the terms on I: $- gr$
multiply the term on L: $- rr = - r^2$
write all the terms as a polynomial and then add or subtract any like terms

$$g^2 + gr - gr - r^2 = g2 - r^2 \blacksquare$$

Both problems in example 2 resulted in binomials. The middle terms were inverses of each other. Inverses combined add up to zero. When the first terms match exactly, and the last terms match exactly, and each binomial has a different sign, two binomials multiplied together always result in a binomial.

EXAMPLE 3:

$(2y - 4q)(8u + 5d)$ multiply the terms on F: $(2y)(8u) = 16uy$
multiply the terms on O: $(2y)(5d) = 10dy$
multiply the terms on I: $(- 4q)(8u) = - 32qu$
multiply the term on L: $(- 4q)(5d) = - 20dq$
write all the terms as a polynomial and then add or subtract any like terms

$16uy + 10dy - 32qu - 20dq$ $\blacksquare$

Two binomials multiplied together with different first and last terms will result in a four term polynomial.

A binomial times a trinomial, or two trinomials multiplying each other cannot use the foil method. For the foil method to work, two terms must multiply two terms. Multiplying a binomial with a trinomial is the same as multiplying 2 terms with 3 terms. Multiplying two trinomials together is the same as multiplying 3 terms by 3 terms. The best way to multiply under these two circumstances is the old fashioned way.

EXAMPLE 4:

```
  734
x 89
 6606    9 times 734
 5872    8 times 734
65326    add the rows together
```
$\blacksquare$

Example 4 was given to remind readers of the old fashioned ways of multiplication. The top number is multiplied by each digit underneath. The same is true for algebra. The top polynomial is multiplied by each term underneath.

EXAMPLE 5:

a)　　$(6v^2 + 4v - 3)\,(6v - 1)$ set up the old fashioned way

$$(6v^2 + 4v - 3)$$
$$\underline{(6v - 1)}$$

$- 6v^2 - 4v + 3$	multiplied top row by -1
$\underline{36v^3 + 24v^2 - 18v}$	multiplied top row by $6v$
$36v^3 + 18v^2 - 22v + 3$	added the two results

b)　　$(8k^2 - 4k - 4)\,(7k^2 + 4k + 5)$ set up the old fashioned way

$$(8k^2 - 4k - 4)$$
$$\underline{(7k^2 + 4k + 5)}$$

$40k^2 - 20k - 20$	multiplied top row by 5
$32k^3 - 16k^2 - 16k$	multiplied top row by $4k$
$\underline{56k^4 - 28k^3 - 28k^2}$	multiplied top row by $7k^2$
$56k^4 + 4k^3 - 4k^2 - 36k - 20$	added the three results ■

Care must be taken when doing problems like example 5. The sign rules from chapter 2 must be followed, and similar terms must be lined up accordingly.

Any combination of polynomials can be multiplied together using the order of operations rule of left to right.

EXAMPLE 6:

$9r\,(8r - 6)\,(7r - 1)$	
$(72r^2 - 54r)(7r - 1)$	distribution property was uses on first multiplication
$504r^3 - 72r^2 - 378r^2 + 54r$	foil method was used
$504r^3 - 450r^2 + 54r$	added like terms ■

There are two forms of multiplication left that bears mentioning. The form of $(a + b)^2$ and $(a - b)^2$ can be multiplied. $(a + b)$ and $(a - b)$ are being squared, and are rewritten $(a + b)\,(a + b)$ and $(a - b)\,(a - b)$. Remember squared means that something multiplies itself. It should be obvious that the foil method can be used to solve these types of problems.

PROBLEMS:

Multiply.

1)　　$(a + 2)\,(a + 6)$　　　　　　　　　　2)　　$(g - 7)\,(g - 2)$

148

3) $(c - 3)(c + 4)$

4) $(d + 7)(d - 2)$

5) $(k - 4)(k + 4)$

6) $(h + g)(h - g)$

7) $(j + 1.1)(j - 1.1)$

8) $(m - \frac{2}{3})(m + \frac{2}{3})$

9) $(7q + 2d)(2q + 3d)$

10) $(3e + 6g)(8e - 2g)$

11) $(14x - 9y)(10x - 11y)$

12) $(1.2k - 1.3m)(1.5k + 9m)$

13) $(3c - 4h)(5f + 6g)$

14) $(8n - 2o)(4j - 3p)$

15) $(a + \frac{1}{2})^2$

16) $(g - 9)^2$

17) $(k + 1.2)(k - 1.3)$

18) $(a - 4.1)(a - 4.2)$

19) $693 \cdot 91$

20) $581 \cdot 72$

21) $809 \cdot 530$

22) $987 \cdot 349$

23) $(a^2 + 2a - 3)(a - 13)$

24) $(s^2 - 3s + 4)(s + 5)$

25) $(6g^2 - 4g - 6)(2g^2 + 2g - 2)$

26) $(a^2 + 2a + 1)(a^2 - 7a + 1)$

27) $a(4a - 6)(6a + 1)$

28) $2c(4c - 5)(2c - 1)$

SECTION 6.4
DIVIDING POLYNOMIALS

Dividing monomials by monomials is exactly the same as the second law of exponents. This type of division will not be repeated here. Binomials and up can be divided by a monomial. This division can be thought of as a distribution under division. Everything in parentheses will be divided by a monomial. Parentheses must be used when dividing by any polynomial. Without parentheses only two terms will be divided. The order of operations takes precedence when parentheses are not involved. Example 1 will show how the parentheses work. We will start doing division with Example 2.

EXAMPLE 1:

a) $14x^2 + 12x \div 2x$ Only $12x$ gets divided.

b) $(14x^2 + 12x) \div 2x$ Every thing in parentheses gets divided. ■

Under the order of operations, everything in parentheses gets divided by 2x. There is no work inside the parentheses; therefore, everything in the parentheses gets divided.

EXAMPLE 2:

a) $(9j - 21) \div 3$ Rewrite with 3 underneath.

$\dfrac{9j}{3} - \dfrac{21}{3}$ Just divide by 3.

$3j - 7$ is the answer.

b) $(14x^2 + 12x) \div 2x$ Rewrite with $2x$ underneath everything to show all divisions.

$\dfrac{14x^2}{2x} + \dfrac{12x}{2x}$ Use 2nd law of exponents.

$7x + 6$ is the answer.

c) $(75h^3 + 55h^2 - 60h) \div 5h$ Rewrite with $5h$ underneath everything to show all divisions.

$\dfrac{75h^3}{5h} + \dfrac{55h^2}{5h} - \dfrac{60h}{5h}$ Use 2nd law of exponents.

$15h^2 + 11h - 12$ is the answer. ■

The next type of division is algebraic long division. Polynomials are divided by at least a binomial. Example 3 will show numerical long division as a reminder on how long division is done. Example 4 will show the algebraic long division.

150

EXAMPLE 3:

a) $12\overline{)3264}$ First, 32 divided by 12 is 2 with leftover. Place 2 on top and then multiply 12 by 2.

$$
\begin{array}{r}
2 \\
12\overline{)3264} \\
24
\end{array}
$$
32 minus 24 is the next step.

$$
\begin{array}{r}
2 \\
12\overline{)3264} \\
24 \\
8
\end{array}
$$
Bring down the 6 and repeat division process.

$$
\begin{array}{r}
2 \\
12\overline{)3264} \\
24 \\
86
\end{array}
$$

86 divided by 12 is 7 with leftover. Place 7 on top after the 2, and then multiply 7 by 12.

$$
\begin{array}{r}
27 \\
12\overline{)3264} \\
24 \\
86 \\
84
\end{array}
$$
86 minus 84.

$$
\begin{array}{r}
27 \\
12\overline{)3264} \\
24 \\
86 \\
84 \\
2
\end{array}
$$
Bring down the 4 and repeat division process.

```
        27
   12)3264
     24
     86
     84
     24
```

24 divided by 12 is two with no leftover. Place 2 on top after 7 and then multiply 2 by 12.

```
       272
   12)3264
     24
     86
     84
      24
      24
```

24 minus 24.

```
       272
   12)3264
     24
     86
     84
      24
      24
       0
```

There is no remainder. The answer is 272.

b)16)3612   First, 36 divided by 16 is 2 with leftover. Place 2 on top and then multiply 16 by 2.

```
       2
   16)3612   36 minus 32.
     32
```

```
      2
16)3612      Bring down the one and repeat division process.
  32
   4
```

```
      2
16)3612
  32
  41
```

41 divided by 16 is 2 with leftover. Place 2 on top after the 2, and then multiply 2 by 16.

```
     22
16)3612
  32
  41   41 minus 32.
  32
```

```
     22
16)3612
  32
  41      Bring down the 2 and repeat division process.
  32
   9
```

```
     22
16)3612
  32
  41
  32
  92
```

92 divided by 16 is 5 and leftover. Place 5 on top after 3, and then multiply 5 by 16.

$$\begin{array}{r} 225 \\ 16\overline{)3612} \\ \underline{32} \\ 41 \\ \underline{32} \\ 92 \\ \underline{80} \end{array}$$ 92 minus 80.

$$\begin{array}{r} 225 \\ 16\overline{)3612} \\ \underline{32} \\ 41 \\ \underline{32} \\ 92 \\ \underline{80} \\ 12 \end{array}$$

There is a remainder of 12. The answer is 225 R 12 ■

Algebraic long division is similar to numerical long division. Each term under the division sign gets divided by the first term that is outside of the division sign. Multiplications and subtractions are the same.

EXAMPLE 4:

a) $(j^2 + 7j + 12) \div (j + 4)$ First set up the old fashioned way.

$j+4\overline{)j^2 + 7j + 12}$ First, j^2 divided by j is j. Place j on top and then multiply j by $(j + 4)$.

$$\begin{array}{r} j \\ j+4\overline{)j^2 + 7j + 12} \\ \underline{j^2 + 4j} \end{array}$$ Subtract down.

$$\begin{array}{r} j \\ j+4\overline{)j^2 + 7j + 12} \\ \underline{j^2 + 4j} \\ 3j \end{array}$$ Bring down + 12 and repeat division process.

154

$$\begin{array}{r} j \\ j+4\overline{)j^2+7j+12} \\ \underline{j^2+4j} \\ 3j+12 \end{array}$$

3j divided by j is 3. Place + 3 on top after j and then multiply 3 by (j + 4).

$$\begin{array}{r} j+3 \\ j+4\overline{)j^2+7j+12} \\ \underline{j^2+4j} \\ 3j+12 \quad \text{Subtract down.} \\ \underline{3j+12} \end{array}$$

$$\begin{array}{r} j+3 \\ j+4\overline{)j^2+7j+12} \\ \underline{j^2+4j} \\ 3j+12 \\ \underline{3j+12} \\ 0 \end{array}$$

There is no remainder. The answer is $j+3$.

b) $(5r^2 - 7r + 31) \div (r - 2)$ First set up the old fashioned way.

$$r-2\overline{)5r^2-7r+31}$$

$5r^2$ divided by r is $5r$. Place $5r$ on top and then multiply by $5r$ by $(r-2)$.

$$\begin{array}{r} 5r \\ r-2\overline{)5r^2-7r+31} \quad \text{Subtract down.} \\ \underline{5r^2-10r} \end{array}$$

$$\begin{array}{r} 5r \\ r-2\overline{)5r^2-7r+31} \quad \text{Bring down} + 31 \text{ and repeat division process.} \\ \underline{5r^2-10r} \\ 3r \end{array}$$

$$\begin{array}{r} 5r \\ r-2\overline{)5r^2-7r+31} \\ \underline{5r^2-10r} \\ 3r+31 \end{array}$$

$3r$ divided by r is 3. Place $+3$ on top after $5r$ and then multiply 3 by $(r-2)$.

$$\begin{array}{r} 5r+3 \\ r-2\overline{)5r^2-7r+31} \\ \underline{5r^2-10r} \\ 3r+31 \text{ Subtract down.} \\ \underline{3r-6} \end{array}$$

$$\begin{array}{r} 5r+3 \\ r-2\overline{)5r^2-7r+31} \\ \underline{5r^2-10r} \\ 3r+31 \\ \underline{3r-6} \\ 37 \end{array}$$

37 is the remainder. The answer is $5r+3+\dfrac{37}{r-2}$.

All remainders must be written as a fraction. The remainder goes on top of the polynomial that is outside of the division symbol.

REMINDER: All negative signs get changed to a positive when subtracted.

EXAMPLE 5:

$(L^3-6L^2+7) \div (L+4)$ First place in missing piece with a 0 coefficient.

$(L^3-6L^2+0L+7) \div (L+4)$ Set up old fashioned way.

$$L+4\overline{)L^3-6L^2+0L+7}$$

L^3 divided by L is L^2. Place L^2 on top and then multiply L^2 by $(L+4)$.

$$\begin{array}{r} L^2 \\ L+4\overline{)L^3-6L^2+0L+7} \text{ Subtract down.} \\ \underline{L^3+4L^2} \end{array}$$

$$\begin{array}{r} L^2 \qquad\qquad \\ L+4\overline{)L^3 - 6L^2 + 0L + 7} \\ \underline{L^3 + 4L^2} \qquad\qquad \\ -10L^2 \qquad\qquad \end{array}$$ Bring down + 0L and repeat division process.

$$\begin{array}{r} L^2 \qquad\qquad \\ L+4\overline{)L^3 - 6L^2 + 0L + 7} \\ \underline{L^3 + 4L^2} \qquad\qquad \\ -10L^2 + 0L \qquad \end{array}$$

$-10L^2$ divided by L is -10L. Place $-10L$ on top after L^2 and then multiply $-10L$ by (L + 4)

$$\begin{array}{r} L^2 - 10L \qquad\qquad \\ L+4\overline{)L^3 - 6L + 0L + 7} \\ \underline{L^3 + 4L^2} \qquad\qquad \\ -10L^2 + 0L \qquad \\ \underline{-10L^2 - 40L} \end{array}$$ Subtract down.

$$\begin{array}{r} L^2 - 10L \qquad\qquad \\ L+4\overline{)L^3 - 6L^2 + 0L + 7} \\ \underline{L^3 + 4L^2} \qquad\qquad \\ -10L^2 + 0L \qquad \\ \underline{-10L^2 - 40L} \\ 40L \qquad \end{array}$$ Bring down +7 and repeat division process.

$$\begin{array}{r} L^2 - 10L \qquad\qquad \\ L+4\overline{)L^3 - 6L^2 + 0L + 7} \\ \underline{L^3 + 4L^2} \qquad\qquad \\ -10L^2 + 0L \qquad \\ \underline{-10L^2 - 40L} \\ 40L + 7 \qquad \end{array}$$

$40L$ divided by L is 40. Place +40 on top after $-10L$ and then multiply 40 by $(L + 4)$.

$$\begin{array}{r} L^2 - 10L + 40 \\ L+4\overline{\smash)L^3 - 6L^2 + 0L + 7} \\ \underline{L^3 + 4L^2} \\ -10L^2 + 0L \\ \underline{-10L^2 - 40L} \qquad \text{Subtract down.} \\ 40L + 7 \\ \underline{40L + 160} \end{array}$$

$$\begin{array}{r} L^2 - 10L + 40 \\ L+4\overline{\smash)L^3 - 6L^2 + 0L + 7} \\ \underline{L^3 + 4L^2} \\ -10L^2 + 0L \\ \underline{-10L^2 - 40L} \\ 40L + 7 \\ \underline{40L + 160} \\ -153 \end{array}$$

- 153 is the remainder. The answer is $L^2 - 10L + 40 - \dfrac{153}{L+4}$. ■

The missing piece $(0L)$ was used as a place holder so that the flow of the problem remains the same.

EXAMPLE 6:

$(6s^2 - 7s + 15) \div (3s - 5)$ First set up old fashioned way.

$3s-5\overline{\smash)6s^2 - 7s + 15}$ First $6s^2$ divided by $3s$ is $2s$. Place $2s$ on top and then multiply $2s$ by $(3s - 5)$.

$$\begin{array}{r} 2s \\ 3s-5\overline{\smash)6s^2 - 7s + 15} \\ \underline{6s^2 - 10s} \end{array}$$ Subtract down.

$$\begin{array}{r} 2s \\ 3s-5\overline{\smash)6s^2 - 7s + 15} \\ \underline{6s^2 - 10s} \\ 3s \end{array}$$ Bring down +15 and repeat division.

158

$$\begin{array}{r} 2s \\ 3s-5{\overline{\smash{\big)}\,6s^2-7s+15}} \\ \underline{6s^2-10s} \\ 3s+15 \end{array}$$

$3s$ divided by $3s$ is 1. Place +1 on top after $2s$, and then multiply $(3s - 5)$ by 1.

$$\begin{array}{r} 2s+1 \\ 3s-5{\overline{\smash{\big)}\,6s^2-7s+15}} \\ \underline{6s^2-10s} \\ 3s+15 \text{Subtract down.} \\ 3s-5 \end{array}$$

$$\begin{array}{r} 2s+1 \\ 3s-5{\overline{\smash{\big)}\,6s^2-7s+15}} \\ \underline{6s^2-10s} \\ 3s+15 \\ \underline{3s-5} \\ 20 \end{array}$$

20 is the remainder. The answer is $2s + 1 + \dfrac{20}{3s-5}$ ■

Notice that none of the sign rules changed from chapter 2 when working downwards in long division. Also, notice that whatever the sign is for a remainder remains the sign for the fraction.

All examples had every polynomial in descending order. Everything lines up perfect when polynomials are in descending order. Change polynomials into descending order whenever they are not in descending order.

An alternative to long division will be given in Section 9.2

PROBLEMS:

1) $(6v^8 + 4v^3 + 8) \div 2$ 2) $(3m^3 - 9m^2) \div 3m$

3) $(10s^8 + 5s^6 - 10s^4) \div 5s^3$ 4) $(91t^9 - 7t^7 - 7t^5) \div 7t^5$

5) $7225 \div 36$ 6) $3127 \div 88$

7) $8943 \div 33$ 8) $38,912 \div 64$

9) $(b^2 + 9b + 14) \div (b + 7)$ 10) $(g^2 + 2g - 35) \div (g - 5)$

11) $(t^2 - 7t + 10) \div (t - 2)$ 12) $(h^2 - 3h - 4) \div (h - 4)$

13) $(x^2 + 6x + 9) \div (x + 3)$ 14) $(s^2 - 5s - 6) \div (s - 6)$

15) $(a^2 - 17a + 66) \div (a - 11)$ 16) $(n^2 - 8n - 9) \div (n - 9)$

17) $(d^2 + 4d + 3) \div (d + 3)$ 18) $(c^2 - 3c - 70) \div (c - 10)$

19) $(m^3 + m^2 - 2) \div (m - 1)$ 20) $(h^4 - h^3 - h + 1) \div (h - 1)$

21) $(6n^2 - n - 10) \div (3n - 5)$ 22) $(4v^2 + 6v - 25) \div (2v + 7)$

23) $(2d^3 - 3d^2 + 4d + 4) \div (2d + 1)$ 24) $(8m^3 - 6m^2 + 2m) \div (4m + 1)$

25) $(2v^2 - 3v - 5) \div (v - 3)$ 26) $(4d^2 - 18d - 15) \div (d - 5)$

27) $(m^4 - 1) \div (m - 1)$ 28) $(v^3 - 8) \div (v - 2)$

Students should study Chapter 9 before attempting the last two problems. Use problems 9 through 28.

29)   What problems could be done using the techniques in Chapter 9? Explain answer.

30)   What problems cannot be done using the techniques in Chapter 9? Explain answer.

CHAPTER 7
RADICALS

SECTION 7.1
INTRODUCTION TO RADICALS

Radicals are the exact opposite of exponents. The radical symbol is $\sqrt{}$ where the number inside the radical is the result of some number being raised to a certain power. The power is found in the cup of the radical. If the cup does not have a number then the power is 2. All powers greater than 2 will be given. We read the powers as a root. Example 1 will show how we read radicals, and example 2 will show some problems dealing with radicals.

EXAMPLE 1:

a) $\sqrt{49}$ is read the square root of 49 or the second root of 49

b) $\sqrt[3]{27}$ is read the cube root of 27 or the third root of 27

c) $\sqrt[4]{16}$ is read the fourth root of 16 ■

EXAMPLE 2:

a) $\sqrt{49} = 7$
    What number squared is 49?

b) $\sqrt[3]{27} = 3$
    What number cubed is 27?

c) $\sqrt[4]{16} = 2$
    What number raised to the fourth power is 16? ■

Any radical problem can be solved on a calculator provided there is radical button on the calculator. Square roots can be found by punching in a number followed by the $\sqrt{}$ button. All other roots use the $\sqrt[x]{y}$ button.

EXAMPLE 3: CALCULATOR USE.

a) $\sqrt{49} = ?$ Press 49 then $\sqrt{}$ button or press $\sqrt{}$ button then 49.
    Answer is 7.

The way of pressing buttons for a square root is dependent upon the calculator.

b) $\sqrt[3]{27}$ = ? Press 3 then the $\sqrt[x]{y}$ button then 27, and then the = button

or press 27 then the $\sqrt[x]{y}$ then 3, and then the = button.

      Answer is 3.

c) $\sqrt[4]{16}$ = ? Press 4 then the $\sqrt[x]{y}$ button then 16, and then the = button

or press 16 then the $\sqrt[x]{y}$ button then 4, and then the = button.

      Answer is 2. ■

Once again, the correct way of pressing buttons to solve a radical problem is dependent upon the calculator. Try both ways to solve a radical problem on a calculator. Whichever way the buttons are pressed that gives the correct answer is the only way to press the buttons on a particular calculator.

All examples shown have integer results. Some problems do not have integer results. To see an example, find the square root of 2 on a calculator. The answer looks something like 1.414213562. These results can be rounded to specified decimal places. For example: the square root of 2 rounded to the nearest tenth would be 1.4.

Here is some extra information on the use of calculators. The calculator button configuration shown in example 3 is not universal. Some calculators do not have the $\sqrt[x]{y}$ button, but roots greater than 2 can still be found on these types of calculators. This alternate button configuration will be discussed in section 10.5.

Fractions and negative numbers can be inside of a radical. Fractions inside of radicals will be discussed in section 7.5. Negative numbers inside of radicals will be discussed in section 10.1

The intent of this chapter is to introduce radicals and their simplest forms. A brief discussion of radicals is needed to prepare students for the section on difference of squares in chapter 8. Most authors introduce radicals near the end of their texts. I've found students understand the concept of difference of squares better when the concept of a square root is discussed prior to learning the concept of difference of squares. The difference of squares is simply a subtraction of squares. Readers should not dwell on difference of squares until the appropriate time. Students can always refer back to this section as well as section 7.3, Simplifying Radicals, because the difference of squares will utilize these concepts.

PROBLEMS:

Put the following radicals into words.

1) $\sqrt{7}$

2) $\sqrt[3]{75}$

3) $\sqrt[5]{59}$

4) $\sqrt{800}$

Find the root.

5) $\sqrt{4}$

6) $\sqrt[3]{64}$

7) $\sqrt[5]{243}$

8) $\sqrt{900}$

9) $\sqrt{256}$

10) $\sqrt[3]{2744}$

11) $\sqrt[4]{4096}$

12) $\sqrt{529}$

Round to the nearest thousandth.

13) $\sqrt{8}$

14) $\sqrt[3]{74}$

15) $\sqrt[5]{347}$

16) $\sqrt{500}$

17) $\sqrt{422}$

18) $\sqrt[3]{2871}$

19) $\sqrt[4]{9290}$

20) $\sqrt{176}$

SECTION 7.2
MULTIPLYING RADICALS (PART I)

Roots must match to multiply radicals. Square roots can multiply square roots. Cube roots can multiply cube roots. Any root above a cube can multiply the same root. Different roots can be multiplied provided that the number is the same underneath the radical. The square root of 2 can multiply the cube root of 2, but this won't be explained until Chapter 10.

I have broken multiplication of radicals into two parts. The first part will show multiplication without simplifying. Many texts have simplification of radicals before multiplication. Students have a problem with taking things apart before putting things together. It has been my experience that students have a better understanding of simplification after they see how things are put together.

Part 2 of multiplying radicals will deal with simplification. This will be done in section 7.6.

EXAMPLE 1:

a) $\sqrt{2}\sqrt{7} = \sqrt{14}$ Just multiply numbers

b) $\sqrt[3]{3}\sqrt[3]{10} = \sqrt[3]{30}$ Just multiply the numbers. ■

There were no multiplications signs in Example 1. Radicals act like variables. Variables next to each other indicate multiplication. Radicals next to each other also indicate multiplication.

PROBLEMS:

Leave all answers in radical form.

1) $\sqrt{6}\sqrt{5}$ 2) $\sqrt{4}\sqrt{2}$

3) $\sqrt{8}\sqrt{7}$ 4) $\sqrt{7}\sqrt{6}$

5) $\sqrt[3]{2}\sqrt[3]{8}$ 6) $\sqrt[3]{9}\sqrt[3]{5}$

7) $\sqrt[5]{3}\sqrt[5]{9}$ 8) $\sqrt[4]{5}\sqrt[4]{4}$

9) $\sqrt{70}\sqrt{55}$ 10) $\sqrt{42}\sqrt{27}$

SECTION 7.3
SIMPLIFYING RADICALS

Simplifying radicals, in its simplest form, means that the root of a number is taken. The other way to simplify a radical is to change the form of the radical without computation. We saw that certain radicals had decimal results in section 7.1. Can these radicals be simplified? The number inside the radical can be factored provided that the number inside the radical can be divided by a perfect power. For example, a number inside of a square root must be divided by a perfect square to be simplified. Example 1 will illustrate simplifying radicals.

EXAMPLE 1:

a) $\sqrt{8}$ 8 can be divided by 4.
The root is 2, and 4 is a perfect square.
$\sqrt{8}$ becomes $\sqrt{4}\sqrt{2}$.
$\sqrt{4}\sqrt{2} = 2\sqrt{2}$ The roots of all perfect powers must be taken.

b) $\sqrt[3]{24}$ 24 can be divided by 8.
The root is 3, and 8 is a perfect cube.
$\sqrt[3]{24}$ becomes $\sqrt[3]{8}\sqrt[3]{3}$.
$\sqrt[3]{8}\sqrt[3]{2} = 2\sqrt[3]{2}$ The roots of all perfect powers must be taken.

c) $\sqrt{6}$ 6 cannot be divided by a perfect square.
$\sqrt{6}$ cannot be simplified.

d) $\sqrt[3]{26}$ 26 cannot be divided by a perfect cube.
$\sqrt[3]{26}$ cannot be simplified.

Radicals with variables can be simplified. A variable with an exponent that is equal to or greater than the root can be simplified. Divide the variable's exponent by the root to simplify the radical. Any remainders are left inside the radical. The reasoning behind this concept will be explained in chapter 10.

EXAMPLE 2:

a) $\sqrt{z^2} = z$. 2 divided by 2 is 1 with no remainder.

b) $\sqrt[3]{j^9} = j^3$. 9 divided by 3 with no remainder.

c) $\sqrt{w^7} = w^3\sqrt{w}$. 7 divided by 2 is 3 with 1 left over.

d) $\sqrt[3]{g^5} = g\sqrt[3]{g^2}$. 5 divided by 3 is 1 with 2 left over. ■

Now let us combine variables and numbers.

EXAMPLE 3:

a) $\sqrt{40v^5} = \sqrt{4}\sqrt{10v^5} = 2v^2\sqrt{10v}$ simplify 40 and do division on exponent.

b) $\sqrt{24d^9h^{10}} = \sqrt{4}\sqrt{6d^9h^{10}} = 2d^4h^5\sqrt{6d}$
Simplify 24 and do division on exponents.

c) $\sqrt[3]{81s^7} = \sqrt[3]{27}\sqrt[3]{3s^7} = 3s^2\sqrt[3]{3s}$ simplify 81 and do division on exponent.

d) $\sqrt[3]{40t^4u^7v^3} = \sqrt[3]{8}\sqrt[3]{5t^4u^7v^3} = 2tu^2v\sqrt[3]{5tu}$ simplify 40 and do division on exponents. ■

PROBLEMS:

Simplify if possible and leave answers in radical form.

1) $\sqrt{68}$ 2) $\sqrt{48}$

3) $\sqrt[3]{16}$ 4) $\sqrt[3]{96}$

5) $\sqrt{10}$ 6) $\sqrt{74}$

7) $\sqrt[3]{28}$ 8) $\sqrt[3]{95}$

9) $\sqrt{h^8}$ 10) $\sqrt{j^{10}}$

11) $\sqrt[3]{b^{81}}$ 12) $\sqrt[3]{h^{12}}$

13) $\sqrt{p^9}$ 14) $\sqrt{r^{17}}$

15) $\sqrt[5]{c^8}$ 16) $\sqrt[4]{h^{22}}$

17) $\sqrt{75h^8}$ 18) $\sqrt{44c^3}$

19) $\sqrt{84o^7p^3}$

20) $\sqrt{32m^8n^9}$

21) $\sqrt[3]{40w^8}$

22) $\sqrt[4]{80d^{10}}$

23) $\sqrt[3]{56m^5n^6}$

24) $\sqrt[5]{96x^8y^5z^6}$

25) $\sqrt{9y^9}$

26) $\sqrt{10a}$

27) $\sqrt[5]{12m^{12}}$

28) $\sqrt[4]{2n^{28}}$

29) $\sqrt{72q^3}$

30) $\sqrt{47m^9}$

31) $\sqrt{800x^4y^9}$

32) $\sqrt{300a}$

33) $\sqrt[3]{400g^{13}}$

34) $\sqrt[4]{480m^9}$

35) $\sqrt[3]{560m^4n^{10}}$

36) $\sqrt[5]{960x^3y^{10}z}$

SECTION 7.4
ADDITION AND SUBTRACTION OF RADICALS

Adding and subtracting radicals is the same as adding and subtracting like terms. Recall from chapter 1 that terms can be added together when the unknowns and the exponents match exactly. Radicals must also match exactly to add or subtract them.

EXAMPLE 1:

a) $3\sqrt{10} + 9\sqrt{10} = 12\sqrt{10}$ add coeficients and keep radicals

b) $3\sqrt{10m} - 2\sqrt{10m} = \sqrt{10m}$ subtract coeficients and keep radicals

c) $7\sqrt{5} - 4\sqrt{5} + 6\sqrt{5} = 9\sqrt{5}$ subtract then add coeficients and keep radicals

d) $3\sqrt{10} - \sqrt{5}$ can' t combine unlike terms ■

Another similarity between radicals and like terms is the invisible 1. An invisible 1 is the coefficient when there is no coefficient present.

EXAMPLE 2:

a) $3\sqrt{10} + \sqrt{10} + 9\sqrt{10} = 13\sqrt{10}$ add coeficients and keep radicals

b) $4\sqrt{10c} - \sqrt{10c} - 2\sqrt{10c} = \sqrt{10c}$ subtract coeficients and keep radicals

c) $7\sqrt[3]{5} - 4\sqrt[3]{5} + 6\sqrt[3]{5} - \sqrt[3]{5} = 8\sqrt[3]{5}$ combine left to right and keep radicals ■

The radicals in the first two examples could not be simplified. The next example will show addition and subtraction of radicals that can be simplified. Add or subtract the radicals, and then simplify the radical.

EXAMPLE 3:

a) $3\sqrt{12} + 9\sqrt{12} = 12\sqrt{12} = 12\sqrt{4}\sqrt{3} = 12 * 2\sqrt{3} = 24\sqrt{3}$

A coefficient in front of a radical implies multiplication. The coefficient of a radical is multiplied by the root of a factored number. 4 was factored from the 12 and the 2 was multiplied to the coefficient of 12.

b) $3\sqrt{18r^2} - 2\sqrt{18r^2} = \sqrt{18r^2} = \sqrt{9r^2}\sqrt{2} = 3r\sqrt{2}$

Division of exponents from section 7.3 was used.

c) $3\sqrt{20s^3} + \sqrt{20s^3} - 2\sqrt{20s^3} = 2\sqrt{20s^3} = 2\sqrt{4}\sqrt{5s^3} = 2 * 2s\sqrt{5s} = 4s\sqrt{5s}$

Division of exponents from section 7.3 was used. ▪

Now, let us combine radicals that can be simplified with radicals that cannot be simplified. Simplify radicals first and then add or subtract to the radicals that cannot be simplified.

EXAMPLE 4:

a) $3\sqrt{10} + 9\sqrt{40} = 3\sqrt{10} + 9\sqrt{4}\sqrt{10} = 3\sqrt{10} + 9 * 2\sqrt{10} = 3\sqrt{10} + 18\sqrt{10}$

$= 21\sqrt{10}$

Simplify radicals and then use order of operations.

b) $3g\sqrt{10g} - 2\sqrt{40g^3} = 3g\sqrt{10g} - 2\sqrt{4}\sqrt{10g^3} = 3g\sqrt{10g} - 2*2g\sqrt{10g}$

$= 3g\sqrt{10g} - 4g\sqrt{10g}$

$= -g\sqrt{10g}$

Notice that sign rules from chapter 2 were used.

c) $7\sqrt{5} - 4\sqrt{3} + 6\sqrt{27} = 7\sqrt{5} - 4\sqrt{3} + 6\sqrt{9}\sqrt{3} = 7\sqrt{5} - 4\sqrt{3} + 6 * 3\sqrt{3}$

$= 7\sqrt{5} - 4\sqrt{3} + 18\sqrt{3}$

$= 7\sqrt{5} + 12\sqrt{3}$

Only like radicals can be combined. ▪

PROBLEMS:

Add or subtract and simplify where possible.

1) $\sqrt{2} + 3\sqrt{2}$

2) $7\sqrt{3} + 8\sqrt{3}$

3) $9\sqrt{2a} - 8\sqrt{2a}$

4) $3\sqrt{3x} - 6\sqrt{3x}$

5) $7\sqrt{7} - 3\sqrt{7} + 2\sqrt{7}$

6) $5\sqrt{10} - 9\sqrt{10} + \sqrt{10}$

7) $5\sqrt{3} - \sqrt{2}$

8) $2\sqrt{7} + 7\sqrt{6}$

9) $9\sqrt{5x} + 15\sqrt{5x} - 6\sqrt{5x}$

10) $3\sqrt{3b} - \sqrt{3b} + 5\sqrt{3b}$

170

11) $\sqrt{7y} + 5\sqrt{7y} + 8\sqrt{7y}$

12) $2\sqrt{5c} + 4\sqrt{5c} + 6\sqrt{5c}$

13) $2\sqrt{10z} - 3\sqrt{10z} - 4\sqrt{10z}$

14) $10\sqrt{7d} - 8\sqrt{7d} - \sqrt{7d}$

15) $\sqrt{12} + 5\sqrt{12}$

16) $2\sqrt{8} + 3\sqrt{8}$

17) $4\sqrt{18b^2} - 6\sqrt{18b^2}$

18) $7\sqrt{20c^4} - 5\sqrt{20c^4}$

19) $5\sqrt{27y^3} + 2\sqrt{27y^3} - 3\sqrt{27y^3}$

20) $9\sqrt{72d^5} - 4\sqrt{72d^5}$

21) $4\sqrt{11} + 5\sqrt{99}$

22) $6\sqrt{13} + 7\sqrt{52}$

23) $6g\sqrt{2g} - \sqrt{18g^3}$

24) $8\sqrt{3h^7} - 2h^3\sqrt{27h}$

25) $2\sqrt{5} - 3\sqrt{3} + 4\sqrt{45}$

26) $5\sqrt{6} - 9\sqrt{7} + 10\sqrt{24}$

27) $\sqrt{50} - 8\sqrt{18} + 9\sqrt{32}$

28) $2\sqrt{27} - 3\sqrt{12} + 4\sqrt{300}$

29) $5\sqrt{20} + 2\sqrt{45} - 9\sqrt{80}$

30) $5\sqrt{28} - 2\sqrt{63} + 4\sqrt{99}$

SECTION 7.5
MULTIPLICATION (PART 2) AND DIVISION OF RADICALS

We saw simple radical multiplication in section 7.3. In this section, the results can be simplified.

EXAMPLE 1:

a) $\sqrt{2}\,\sqrt{6} = \sqrt{12} = \sqrt{4}\sqrt{3} = 2\sqrt{3}$ Multiply then simplify

b) $\sqrt{3m}\,\sqrt{8m} = \sqrt{24m^2} = \sqrt{4m^2}\,\sqrt{6} = 2m\sqrt{6}$ Multiply then simplify

c) $\sqrt[3]{3}\sqrt[3]{16} = \sqrt[3]{48} = \sqrt[3]{8}\sqrt[3]{6} = 2\sqrt[3]{6}$ Multiply then simplify. ■

Now let us see how multiplication with coefficients works.

EXAMPLE 2:

a) $8\sqrt{2} * \sqrt{6} = 8\sqrt{12} = 8\sqrt{4}\sqrt{3} = 8 * 2\sqrt{3} = 16\sqrt{3}$ Multiply then simplify

b) $6\sqrt{3m} * 3\sqrt{8m} = 18\sqrt{24m^2} = 18\sqrt{4m^2}\,\sqrt{6} = 18 * 2m\sqrt{6} = 36m\sqrt{6}$ Multiply then simplify

c) $4\sqrt[3]{3} * 6\sqrt[3]{16} = 24\sqrt[3]{48} = 24\sqrt[3]{8}\sqrt[3]{6} = 24 * 2\sqrt[3]{6} = 48\sqrt[3]{6}$ Multiply then simplify.
■

The distribution property and foil method also work with radicals.

EXAMPLE 3:

a) $\sqrt{2}\left(\sqrt{6} + \sqrt{3}\right) = \sqrt{12} + \sqrt{6} = \sqrt{4}\sqrt{3} + \sqrt{6} = 2\sqrt{3} + \sqrt{6}$ Distribute then simplify

b) $6\sqrt{2}\left(6\sqrt{3m} - 3\sqrt{8m}\right) = 36\sqrt{6m} - 18\sqrt{16m} = 36\sqrt{6m} - 18 * 4\sqrt{m} = 36\sqrt{6m} - 72\sqrt{m}$
Distribute then simplify. ■

EXAMPLE 4:

a) $\left(\sqrt{6} + \sqrt{3}\right)\left(\sqrt{6} + \sqrt{7}\right) = \sqrt{36} + \sqrt{42} + \sqrt{18} + \sqrt{21} = 6 + \sqrt{42} + \sqrt{9}\sqrt{2} + \sqrt{21} = 6 + \sqrt{42} + 3\sqrt{2} + \sqrt{21}$
Use foil method first, then simplify

b) $\left(\sqrt{3} - \sqrt{8}\right)\left(\sqrt{3} + \sqrt{8}\right) = \sqrt{9} + \sqrt{24} - \sqrt{24} - \sqrt{64} = 3 - 8 = -5$

   Use foil method first. Simplify the radicals and then combine like terms. ∎

Roots must match to divide radicals. Square roots can divide square roots. Division of radicals is a two step process. First, divide the numbers inside the radicals, and then simplify the radical if it is possible. This division can only be accomplished when the top radical is larger than the bottom one. Division problems where the bottom is larger will be explained in chapter 10.

EXAMPLE 5:

a) $\dfrac{\sqrt{14}}{\sqrt{2}} = \sqrt{7}$ Just divide the numbers. Can' t be simplified.

b) $\dfrac{\sqrt{18}}{\sqrt{2}} = \sqrt{9} = 3$ Divide and simplify.

c) $\dfrac{\sqrt{24}}{\sqrt{3}} = \sqrt{8} = \sqrt{4}\sqrt{2} = 2\sqrt{2}$ Divide and simplify.

d) $\dfrac{\sqrt{32m^3}}{\sqrt{2m}} = \sqrt{16m^2} = 4m$ Divide and simplify. ∎

Now let us see how division with coefficients works.

EXAMPLE 6:

a) $\dfrac{9\sqrt{14}}{3\sqrt{2}} = 3\sqrt{7}$ Just divide the numbers. Can' t be simplified.

b) $\dfrac{8\sqrt{18}}{2\sqrt{2}} = 4\sqrt{9} = 4 * 3 = 12$ Divide and simplify.

c) $\dfrac{70\sqrt{24}}{5\sqrt{3}} = 14\sqrt{8} = 14\sqrt{4}\sqrt{2} = 14 * 2\sqrt{2} = 28\sqrt{2}$ Divide and simplify.

d) $\dfrac{10\sqrt{32m^3}}{5\sqrt{2m}} = 2\sqrt{16m^2} = 2 * 4m = 8m$ Divide and simplify. ∎

PROBLEMS:

Multiply or divide, and then simplify if possible.

1) $\sqrt{2}\,\sqrt{6}$

2) $\sqrt{8}\,\sqrt{3}$

3) $\sqrt{5a}\,\sqrt{10a}$

4) $\sqrt{7b^2}\,\sqrt{14b^2}$

5) $\sqrt[3]{4}\,\sqrt[3]{12}$

6) $\sqrt[3]{9c^2}\,\sqrt[3]{3c}$

7) $2\sqrt{5} \cdot \sqrt{8}$

8) $8\sqrt{7} \cdot 10\sqrt{12}$

9) $4\sqrt{7d} \cdot 8\sqrt{14d}$

10) $2\sqrt{11h^7} \cdot 13\sqrt{18h^9}$

11) $2\sqrt[3]{6} \cdot \sqrt[3]{16}$

12) $8\sqrt[3]{3g} \cdot 5\sqrt[3]{9g^8}$

13) $\sqrt{5}(\sqrt{8}+\sqrt{10})$

14) $\sqrt{7}(\sqrt{11}-\sqrt{12})$

15) $2\sqrt{11}(\sqrt{13h}+\sqrt{12h})$

16) $3\sqrt{13j}(\sqrt{10}+\sqrt{14j})$

17) $(\sqrt{2}+\sqrt{3})(\sqrt{2}+\sqrt{5})$

18) $(\sqrt{6}+\sqrt{7})(\sqrt{6}-\sqrt{8})$

19) $(\sqrt{6}-\sqrt{7})(\sqrt{6}+\sqrt{7})$

20) $(\sqrt{10}+\sqrt{11})(\sqrt{10}-\sqrt{11})$

21) $\sqrt{8}\div\sqrt{2}$

22) $\sqrt{12}\div\sqrt{6}$

23) $\sqrt{40}\div\sqrt{5}$

24) $\sqrt{18}\div\sqrt{2}$

25) $\sqrt{96}\div\sqrt{6}$

26) $\sqrt{88}\div\sqrt{22}$

27) $\sqrt{80k^3}\div\sqrt{5k}$

28) $\sqrt{90m^7}\div\sqrt{2m^5}$

29) $14\sqrt{6}\div7\sqrt{3}$

30) $9\sqrt{10}\div3\sqrt{5}$

31) $10\sqrt{50}\div5\sqrt{5}$

32) $36\sqrt{54}\div4\sqrt{6}$

33) $70\sqrt{65n^3}\div5\sqrt{5n}$

34) $10\sqrt{15o^5}\div2\sqrt{3o^3}$

CHAPTER 8
FACTORING POLYNOMIALS

SECTION 8.1
INTRODUCTION TO FACTORING

Let's go back and take a look at numerical factoring before looking at algebraic factoring. Numerical factoring is breaking down a given integer into integers that divide into the given integer. Every integer has at least two factors. One and the given integer are the two factors for all integers.

EXAMPLE 1:

Factor all integers

a)   The factors of 16 are 1, 2, 4, 8, and 16.

b)   The factors of 86 are 1,2,43, and 86.

c)   The factors of 395 are 1,5,79,395.

d)   The factors of 1250 are 1, 2, 5,10,25,50,125,250,625, and1250. ∎

The last digit is a clue to factoring integers. Any number that has an even last digit can be divided by 2. Any number that ends in a 5 or 0 can be divided by 5. Any number that ends in 0 can be divided by 10. These clues are a starting point to factoring numbers.

Algebraic factoring is breaking down polynomials into other polynomials that divide into the original polynomial. This chapter deals with three types of algebraic factoring. Common factors, difference of squares, and trinomial factoring will be discussed in the first four sections of this chapter. This section will only deal with common factors.

A common factor is a factor that divides into every part of a polynomial. A polynomial has a common factor when every term has the same unknown in it, or has coefficients that can be divided by the same integer.

EXAMPLE 2:

$5x + 15$ has a common factor of 5. Both terms can be divided by 5. Divide 5 into $5x + 15$, and then write the result as a product of the result and 5.

$(5x + 15) \div 5 = x + 3$ Use method from chapter 6.

The factors are written as $5(x + 3)$. The distribution rule can be used to verify that 5 and $(x + 3)$ are factors of $5x + 15$. ∎

EXAMPLE 3:

$x + xy$ has a common factor of x. x divides into both terms. To write out the factors, divide $x + xy$ by x, and then write the result as a product of the result and x.

$(x + xy) \div x = 1 + y$ Use method from chapter 6.

The factors are written as $x(1 + y)$. The distribution rule can be used to verify that x and $(1 + y)$ are factors of $x + xy$. ■

Common factors can occur when the common unknown has a different exponent in each term. In this case, the unknown with the lowest exponent is the common factor. Only the unknown with the lowest exponent can divide into each term.

EXAMPLE 4:

$x^3 + x^2 - x$ has a common factor of x. x divides into all three terms. To write out the factors, divide $x^3 + x^2 - x$ by x, and then write the result as a product of the result and x.

$(x^3 + x^2 - x) \div x = x^2 + x - 1$ Use method from chapter 6.

The factors are written as $x(x^2 + x - 1)$. The distribution rule can be used to verify that x and $(x^2 + x - 1)$ are factors of $x^3 + x^2 - x$. ■

Let us now look at polynomials that have integers and unknowns as common factors.

EXAMPLE 5:

a) $6y^3 + 8y^2$ has a common factor of $2y^2$. $2y^2$ divides into both terms. To write out the factors, divide $6y^3 + 8y^2$ by $2y^2$, and then write the result as a product of the result and $2y^2$.

$(6y^3 + 8y^2) \div 2y^2 = 3y + 4$ Use method from chapter 6.

The factors are written as $2y^2(3y + 4)$. The distribution rule can be used to verify that $2y^2$ and $(3y + 4)$ are factors of $6y^3 + 8y^2$.

b) $9w^3 + 18w^2 - 36w$ has a common factor of $9w$. $9w$ divides into all three terms. To write out the factors, divide $9w^3 + 18w^2 - 36w$ by $9w$, and then write the result as a product of the result and $9w$.

$(9w^3 + 18w^2 - 36w) \div 9w = w^2 + 2w - 4$ Use method from chapter 6.

The factors are written as $9w (w^2 + 2w - 4)$. The distribution rule can be used to verify that $9w$ and $(w^2 + 2w - 4)$ are factors of $9w^3 + 18w^2 - 36w$.

c) $60w^3x^2 - 65w^2x + 80wx^3 - 55wx$ has a common factor of $5wx$. $5wx$ divides into all four terms. To write out the factors, divide $60w^3x^2 - 65w^2x + 80wx^3 - 55wx$ by $5wx$, and then write the result as a product of the result and $5wx$.

$(60w^3x^2 - 65w^2x + 80wx^3 - 55wx) \div 5wx = 12w^2x - 13w + 16x^2 - 11$ Use method from chapter 6.

The factors are written as $5wx (12w^2x - 13w + 16x^2 - 11)$. The distribution rule can be used to verify that $5wx$ and $(12w^2x - 13w + 16x^2 - 11)$ are factors of $60w^3x^2 - 65w^2x + 80wx^3 - 55wx$. ■

PROBLEMS:

Factor completely.

1)   24

2)   45

3)   508

4)   9250

5) $7h + 63$

6) $2g - 20$

7) $5y + 55$

8) $6q - 672$

9) $q + qr$

10) $b - ab$

11) $st + t$

12) $yz - z$

13) $u^3 + u^2 - u$

14) $n^4 + n^3 - n^2 + n$

15) $p^5 - p^4 + p^3 - p^2 + p$

16) $k^6 - k^4 + k^2$

17) $4b^4 + 6b$

18) $3d^3 - 15d^2 + 33d$

19) $24p^3q^2 - 33p^2q + 96pq^3 - 12pq$

20) $60x^3y^5 + 6x^2y^4 - 8x^2y^6 + 2xy$

SECTION 8.2
FACTORING THE DIFFERENCE OF SQUARES

The difference of squares describes a specific binomial. This binomial has a minus sign between two perfect squares. One of the squares must be an unknown. A minus sign that is between two numerical perfect squares is not a binomial but a computational problem. $x^2 - 4$ is a difference of squares.

The formula $a^2 - b^2 = (a + b)(a - b)$ or $a^2 - b^2 = (a - b)(a + b)$ must be used to factor differences of squares. Multiplication is commutative and the order of the factors does not matter. The steps to factoring a difference of squares are as follows:

Step 1: Take the square root from both sides of the minus sign
Step 2: Place both roots in their appropriate place in the factored form which is on the right side of the equal sign in the formula.

The two factors of a difference of squares must have different signs. Remember from chapter 6 that $(a + b)(a - b) = a^2 - b^2$ after using the FOIL method.

EXAMPLE 1:

Factor: $o^2 - 81$
Step 1: The square root of o^2 is o, and the square root of 81 is 9
Step 2: Place both roots in their appropriate place in the factored form which is on the right side of the equal sign in the formula. o is a and 9 is b.

The solution is $(o + 9)(o - 9)$ ∎

EXAMPLE 2:

Factor: $9a^2 - 16$
Step 1: The square root of $9a^2$ is $3a$, and the square root of 16 is 4.
Step 2: Place both roots in their appropriate place in the factored form which is on the right side of the equal sign in the formula. $3a$ is a, and 4 is b.

The solution is $(3a - 4)(3a + 4)$ ∎

EXAMPLE 3:

Factor: $36x^2 - 25y^2$
Step 1: The square root of $36x^2$ is $6x$, and the square root of $25y^2$ is $5y$.
Step 2: Place both roots in their appropriate place in the factored form which is on the right side of the equal sign in the formula. $6x$ is a, and $5y$ is b.

The solution is $(6x + 5y) (6x - 5y)$ ■

EXAMPLE 4:

Factor: $64 - h^4$

Step 1: The square root of 64 is 8, and the square root of h^4 is h^2.

Step 2: Place both roots in their appropriate place in the factored form which is on the right side of the equal sign in the formula. 8 is a, and h^2 is b.

The solution is $(8 + h^2) (8 - h^2)$ ■

CAUTION: Many students tend to change the order of problems like EXAMPLE 4. Most students would rewrite EXAMPLE 4 as $h^4 - 64$, and claim the answer is $(h^2 + 8)(h^2 - 8)$. $(h^2 + 8)(h^2 - 8) = h^4 - 64$, and $(8 + h^2) (8 - h^2) = 64 - h^4$. It is clear that the solutions do not match.

Now let us look at an example that uses factoring methods from the previous section, and this one.

EXAMPLE 5:

Factor: $27n^3 - 3n$

There is a common factor of $3n$

$27n^3 - 3n = 3n (9n^2 - 1)$ using the method from the last section.

There is a difference of squares that needs to be factored.

Step 1: The square root of $9n^2$ is $3n$, and the square root of 1 is 1.

Step 2: Place both roots in their appropriate place in the factored form which is on the right side of the equal sign in the formula. $3n$ is a, and 1 is b.

The solution is $3n (3n + 1) (3n - 1)$

The $3n$ that was factored from the original binomial must be in the solution. It is left up to the reader to multiply the solution to get the original binomial. ■

Factoring is not completed until all factors are found as was the case in EXAMPLE 5.

SUMMARY: Factor out a common factor if there is one.
Take the square root of both terms.
Place factors in appropriate places using formula.

PROBLEMS:

Factor completely.

1) $a^2 - 49$

2) $b^2 - 4$

3) $c^2 - 1$

4) $d^2 - 484$

5) $9h^2 - 36$

6) $16g^2 - 25$

7) $81h^2 - 64$

8) $100j^2 - 121$

9) $49m^2 - 100n^2$

10) $81o^2 - 144p^2$

11) $36q^2 - 121r^2$

12) $64s^2 - 169g^2$

13) $25 - u^2$

14) $49 - v^4$

15) $16 - w^6$

16) $36 - x^2$

17) $18y^3 - 2y$

18) $12z^3 - 75z$

19) $48a^2 - 27a^4$

20) $50b - 2b^3$

21) $c^4 - 100$

22) $d^2h^2 - 196$

23) $9 - e^2g^2$

24) $h^4 - 225j^2$

25) $18k^3 - 2km^2$

26) $12n^3o - 75no^3$

SECTION 8.3
FACTORING TRINOMIALS PART 1

Trinomials that only have a coefficient of 1 in the first term will be discussed in this section. $x^2 - x + 1$ is an example of the type of trinomials that will be factored in this section. The key to factoring trinomials in this section is the sign of the third term. Factors have the sign from the second term if the trinomial's third term has a plus sign. Factors have one plus sign and one negative sign if the trinomial's third term has a minus sign. Also, the third term must be broken down into integers that make up the middle coefficient. Examples 1 through 4 will show how the third term is broken down.

Recall from the FOIL method that the L multiplication generates the third term. The third term was either positive or negative depending on whether or not that the L terms had the same sign or not.

EXAMPLE 1:

Factor: $g^2 + 9g + 20$

First break down the third term in pairs that multiply together to make 20
20, 1   10, 2   5, 4

Since the third term is positive, which pair adds up to 9? 5, 4 will be the pair used.

$(g + 5)(g + 4)$ is the solution. ■

Notice that the first term was broken down in the same manner as a difference of squares. The similarity only occurs in part 1 of factoring trinomials.

EXAMPLE 2:

Factor: $w^2 - 16w + 15$

First break down the third term in pairs that multiply together to make 15
15, 1   5, 3

Since the third term is positive, which pair adds up to 16? 15, 1 will be the pair used.

$(w - 15)(w - 1)$ is the solution. ■

Notice the sign from the second term was kept in both examples because the third term was positive.

Now we will see what happens when the third term is negative. A negative sign is the opposite of a positive sign; therefore, we will use the integers that subtract to make the middle coefficient from the pairs from the third term.

EXAMPLE 3:

Factor: $c^2 + 8c - 20$

First break down the third term in pairs that multiply together to make 20
20, 1   10, 2   5, 4

Since the third term is negative, which pair subtracts to 8? 10, 2 will be the pair used.

$(c + 10)(c - 2)$ is the solution. ■

The sign from the middle is usually placed on the largest of the pair used.

The OI multiplication from FOIL can be used to check to see if the sign placement is correct. $-2c + 10c = 8c$. Our sign placement is correct.

EXAMPLE 4:

Factor: $m^2 - 9m - 36$

First break down the third term in pairs that multiply together to make 36
36, 1   18, 2   12, 3   9, 4   6, 6

Since the third term is negative, which pair subtracts to 9? 12, 3 will be the pair used.

$(m - 12)(m + 3)$ is the solution.

OI check: $3m - 12m = -9m$ ■

EXAMPLE 5:

Factor: $x^2 - x + 1$

Break down the third term into pairs that multiply together to make 1.
1, 1

Since the third term is positive, which pair adds up to 1? The pair cannot be used because the sum of the pair does not add up to 1.
There is no solution. ■

Whenever a pair from the third term never adds or subtracts to make the middle coefficient, the answer is no solution. Chapter 10 will explain what happens when these types of trinomials are present.

Now let us look at an example that uses factoring methods from the first section, and this one.

EXAMPLE 6:

Factor: $7a^2 + 14a - 21$

Factor out the common 7: $7(a^2 + 2a - 3)$

Now factor: $(a^2 + 2a - 3)$

First break down the third term in pairs that multiply together to make 3
3, 1

Since the third term is negative, which pair subtracts to 2? 3, 1 will be the pair used.

$7(a + 3)(a - 1)$ is the solution.

OI check: $-a + 3a = 2a$ ■

All solutions can be verified by multiplication.

> **SUMMARY:** Factor out a common factor if there is one.
> Breakdown down third term into pairs
> Find pair that adds or subtracts to make middle coefficient
> Place factors in appropriate places with appropriate signs.

PROBLEMS:

Factor completely. Write not factorable if trinomial cannot be factored.

1) $p^2 + 6p + 8$

2) $q^2 + 8q + 15$

3) $r^2 + 13r + 30$

4) $s^2 + 14s + 33$

5) $u^2 - 9u + 20$

6) $v^2 - 11v + 28$

7) $w^2 - 8w + 15$

8) $y^2 - 14y + 49$

9) $z^2 + z - 6$

10) $a^2 + 2a - 15$

11) $b^2 + 3b - 10$

12) $c^2 + 6c - 55$

13) $d^2 - 2d - 63$

14) $h^2 - h - 20$

15) $g^2 - 6g - 40$

16) $h^2 - 10h - 24$

17) $j^2 + j + 2$

18) $k^2 - 3d + 7$

19) $m^2 + 4m + 5$

20) $o^2 - 12o - 35$

21) $2p^2 + 16p + 30$

22) $8u^2 - 64u + 56$

23) $9b^2 + 9b - 180$

24) $3h^2 - 6h - 297$

25) $j^2 + 10j + 25$

26) $6k^2 - 48k - 120$

27) $2m^6 - 8m^5 - 42m^4$

28) Could the answer of #8 be written in a different manner? Refer to section 6.3.

29) Could the answer of #25 be written in a different manner? Refer to section 6.3.

30) Is $(n + 10)(n - 9)$ the same as $(n - 9)(n + 10)$? Refer to section 8.2.

SECTION 8.4
FACTORING TRINOMIALS BY GROUPING

  Factoring a trinomial by grouping is taking a trinomial and turning it into a four term polynomial, and then grouping the first two terms together, and the last two terms together. There are five steps to grouping. First, multiply the first coefficient and the third term together. Secondly, rewrite the product of the first coefficient, and the third term as a sum or difference, with the unknown that will make the middle term. Place parentheses around the first two terms, and the last two terms is the third step. The fourth step is to factor out the common factor from each of the parentheses. Finally, place the binomial inside the parentheses next to a binomial that is made up of the terms that are on the outside of the parentheses. If the procedure is not clear to the reader then it should become apparent following the examples.

  We will use the examples from the previous section because grouping is an alternative to factoring part 1 trinomials. This method will be applied to factoring part 2 trinomials that will be discussed in the next section.

EXAMPLE 1:

  Factor: $g^2 + 9g + 20$

  First multiply the first coefficient and the third term.
  $1 \cdot 20 = 20$

  Rewrite 20 as a sum of 9 with g
  $g^2 + 5g + 4g + 20$ Notice that the middle terms are equivalent as the original middle term. Don't combine the middle terms.

  Place parentheses around the first two terms and the last two terms and then factor out a common factor from each group of terms.
  $(g^2 + 5g) + (4g + 20)$

  $g(g + 5) + 4(g + 5)$

  $(g + 5)(g + 4)$ is the solution. ■

EXAMPLE 2:

  Factor: $c^2 + 8c - 20$

  First multiply the first coefficient and the third term.
  $1(-20) = -20$

Rewrite 20 as a difference of 8 with c
$c^2 - 10c + 2c - 20$. Notice that the middle terms are equivalent as the original middle term. Don't combine the middle terms. Also, place the negative term second so that no signs will need to be changed. Changing signs will be shown in a later example.

Place parentheses around the first two terms and the last two terms and then factor out common factors from each group of terms.
$(c^2 - 10c) + (2c - 20)$

$c(c - 10) + 2(c - 10)$

$(c - 10)(c + 2)$ is the solution. ■

EXAMPLE 3:

Factor: $m^2 - 9m - 36$

First multiply the first coefficient and the third term.
$1(-36) = -36$

Rewrite -36 as a difference of -9 with m
$m^2 - 12m + 3m - 36$

Place parentheses around the first two terms and the last two terms and then factor out common factors from each group of terms.
$(m^2 - 12m) + (3m - 36)$

$m(m - 12) + 3(m - 12)$

$(m - 12)(m + 3)$ is the solution. ■

Now let us look at examples that have a sign change.

EXAMPLE 4:

Factor: $w^2 - 16w + 15$

First multiply the first coefficient and the third term.
$1 \cdot 15 = 15$

Rewrite 15 as a sum of 16 with w. Remember to keep the signs since the third term is positive.

$w^2 - 15w - w + 15$
Place parentheses around the first two terms and the last two terms and then factor out common factors from each group of terms.

$(w^2 - 15w) - (w - 15)$ Notice that when the parentheses separates a negative sign and the unknown the other sign changes.

$w(w - 15) - 1(w - 15)$ When no common factor is apparent then the common factor is 1.

$(w - 15)(w - 1)$ is the solution. ■

All trinomials similar to example 4 have a sign change. By placing the parenthesis between the w and the negative sign, the w becomes a positive; therefore, the sign of the last term must also change.

EXAMPLE 5:

Factor: $x^2 - x + 1$

First multiply the first coefficient and the third term.
$1 \cdot 1 = 1$

Rewrite 1 as a sum of 1 with x. Remember to keep the signs since the third term is positive.

$x^2 - x - 0x + 1$

Place parentheses around the first two terms and the last two terms and then factor out common factors from each group of terms.
$(x^2 - x) - (0x - 1)$ Sign changes the same as example 4

$x(x - 1) - 1(0x - 1)$

There is no solution. Parentheses must match for a solution to occur. ■

Now let us look at an example that uses factoring methods from the first section, and this one.

EXAMPLE 6:

Factor: $7a^2 + 14a - 21$

Factor out the common 7: $7(a^2 + 2a - 3)$

Now factor: $(a^2 + 2a - 3)$

First multiply the first coefficient and the third term.
$1(-3) = -3$

Rewrite -3 as a difference of 2 with a

$a^2 - a + 3a - 3$

Place parentheses around the first two terms and the last two terms and then factor out common factors from each group of terms.
$(a^2 - a) + (3a - 3)$

$a(a - 1) + 3(a - 1)$

$7(a - 1)(a + 3)$ is the solution. ■

All solutions can be verified by multiplication.

SUMMARY: Multiply first coefficient and third term

Rewrite product of first coefficient and third term with the unknown as a sum or difference that makes up the original second term

Place parentheses around the first two terms and the last two terms and then factor out common factors from each group of terms.

Write solution as a product of the parentheses and a binomial of what is outside the parentheses.

PROBLEMS:

Factor completely. Write not factorable if trinomial cannot be factored.

1) $p^2 + 4p + 3$ 2) $q^2 - 2q - 3$

3) $r^2 - 12r + 35$ 4) $s^2 + 5s - 24$

5) $u^2 - 9u + 14$ 6) $v^2 + 10v + 25$

7) $w^2 - 2w - 63$ 8) $y^2 - 14y + 49$

9) $z^2 + 4z - 77$ 10) $a^2 + 6a + 5$

11) $b^2 - 8b + 15$ 12) $c^2 - 10c + 21$

13) $d^2 + 12d + 27$ 14) $h^2 - 11h + 30$

15) $g^2 + 8g + 16$ 16) $h^2 - 2h - 48$

17) $j^2 - 30j + 81$ 18) $k^2 + 8k - 33$

19) $m^2 + 2m - 3$ 20) $o^2 + 10o + 21$

21) $2p^2 - 8p - 10$ 22) $3u^2 - 12u + 9$

23) $3b^2 - 18b + 15$ 24) $2h^2 - 2h - 84$

25) $j^2 + 6j - 72$ 26) $3k^2 - 12k - 96$

27) $2m^6 - 10m^5 - 100m^4$

Refer to section 6.3.

28)   Could the answer of #6 be written in a different manner?

29)   Could the answer of #8 be written in a different manner?

30)   Could the answer of #15 be written in a different manner?

SECTION 8.5
FACTORING TRINOMIALS PART 2

Trinomials that have a coefficient larger than 1 in the first term are the topic of discussion in this section. This coefficient can also occur after a common factor is factored out. The grouping method will be used to factor these trinomials.

EXAMPLE 1:

Factor: $5c^2 + 18c + 9$

First multiply the first coefficient and the third term.
$5 \cdot 9 = 45$

Rewrite 45 as a sum of 18 with c
$5c^2 + 15c + 3c + 9$

Place parentheses around the first two terms and the last two terms and then factor out common factors from each group of terms.
$(5c^2 + 15c) + (3c + 9)$

$5c(c + 3) + 3(c + 3)$

$(c + 3)(5c + 3)$ is the solution. ■

EXAMPLE 2:

Factor: $4k^2 + 3k - 10$

First multiply the first coefficient and the third term.
$4(-10) = -40$

Rewrite -40 as a difference of 3 with k
$4k^2 - 5k + 8k - 10$

Place parentheses around the first two terms and the last two terms and then factor out common factors from each group of terms.
$(4k^2 - 5k) + (8k - 10)$

$k(4k - 5) + 2(4k - 5)$

$(4k - 5)(k + 2)$ is the solution. ■

EXAMPLE 3:

Factor: $2k^2 - k - 1$
First multiply the first coefficient and the third coefficient.
$2(-1) = -2$

Rewrite -2 as a difference of -1 with k.
$2k^2 - 2k + k - 1$

Place parentheses around the first two terms and the last two terms and then factor out common factors from each group of terms.
$(2k^2 - 2k) + (k - 1)$

$2k(k - 1) + (k - 1)$

A common factor of 1 is always present when it looks like there is no common factor.

$(k - 1)(2k + 1)$ is the solution. ■

EXAMPLE 4:

Factor: $5p^2 - 4pq - 12q^2$
First multiply the first coefficient and the third coefficient.
$5(-12) = -60$

Rewrite -60 as a difference of -4 with pq
$5p^2 - 10pq + 6pq - 12q^2$

Place parentheses around the first two terms and the last two terms and then factor out common factors from each group of terms.
$(5p^2 - 10pq) + (6pq - 12q^2)$

$5p(p - 2q) + 6q(p - 2q)$

$(p - 2q)(5p + 6q)$ is the solution. ■

Notice that there was not any change in the procedure even though there were two unknowns present.

EXAMPLE 5:

Factor: $3c^2 - 14c + 8$

First multiply the first coefficient and the third term.
$3 \cdot 8 = 24$

Rewrite 24 as a sum of 14 with c. Remember to keep the signs since the third term is positive.
$3c^2 - 2c - 12c + 8$

Place parentheses around the first two terms and the last two terms and then factor out common factors from each group of terms. Watch the sign change.
$(3c^2 - 2c) - (12c - 8)$

$c(3c - 2) - 4(3c - 2)$

$(3c - 2)(c - 4)$ is the solution. ■

EXAMPLE 6:

Factor: $9p^2 + 4p - 6$

First multiply the first coefficient and the third term.
$9(-6) = -54$

Rewrite -54 as a difference of 4 with p.

-54 cannot be rewritten as a difference of 4.

Whenever a product of the first coefficient and the third term cannot make a sum or difference of the original middle term then there is no solution. ■

Now let us look at an example that uses factoring methods from the first section, and this one.

EXAMPLE 7:

Factor: $4o^2 + 6o - 10$

Factor out the common 2: $2(2o^2 + 3o - 5)$

First multiply the first coefficient and the third term.
$2(-5) = -10$

Rewrite -10 as a difference of 3 with o
$2o^2 - 2o + 5o - 5$

Place parentheses around the first two terms and the last two terms and then factor out common factors from each group of terms.

$(2o^2 - 2o) + (5o - 5)$

$2o(o - 1) + 5(o - 1)$

$2(o - 1)(2o + 5)$ is the solution. ▪

The 2 that was pulled out in the beginning of Example 7 was a common factor. It must be placed in front of the two binomial factors from the resulting trinomial. A common factor is still a factor from the original trinomial, and must be written as part of the solution. All solutions can be verified by multiplication.

SUMMARY: Multiply first coefficient and third term

Rewrite product of first coefficient and third term with the unknown as a sum or difference that makes up the original second term

Place parentheses around the first two terms and the last two terms and then factor out common factors from each group of terms.

Write solution as a product of the parentheses and a binomial of what is outside the parentheses.

HINT: Always factor out a common factor first. Part 2 trinomials are easier to factor with the common factored removed.

PROBLEMS:

Factor completely. Write not factorable if trinomial cannot be factored.

1) $2p^2 + 5p - 3$ 2) $6q^2 - 19q + 10$

3) $3r^2 - 14r - 5$ 4) $6s^2 + 19s + 19$

5) $6u^2 - 7u - 3$ 6) $15v^2 + v - 6$

7) $6w^2 + 25w - 25$ 8) $9y^2 - 12y + 4$

9) $12z^2 - 8z - 15$ 10) $3a^2 + 7a - 6$

11) $8b^2 - 27b - 20$ 12) $2c^2 + 3c + 1$

13) $5d^2 - 8d - 4$ 14) $9h^2 + 4h - 5$

15) $6g^2 - 17g + 12$

16) $36h^2 - 3h - 5$

17) $20j^2 - 20j - 15$

18) $12k^2 + 19k + 4$

19) $8m^2 - 6m - 9$

20) $20o^2 - 33o + 10$

21) $16p^2 + 40p + 25$

22) $12u^2 + 11u - 15$

23) $24b^2 + 5b - 36$

24) $24h^2 - 18h - 6$

25) $9p^2 + 30pq + 21q$

26) $12u^2 + 2uv - 24v^2$

27) $4b^2 + 13bc + 3c^2$

28) $6j^2 - 13jk - 5k^2$

29) $2k^2 + 7k + 3$

30) $9d^2 - 30d + 25$

31) $10h^2 - 7h - 12$

32) $8g^2 + 14g - 15$

33) $81x^2 - 36xy + 4y^2$

34) $6x^2 - 5xy - 56y^2$

35) $3h^2 + 16h + 5$

36) $25j^2 - 20j + 4$

37) $2u^2 + 13uv - 7v^2$

38) $3x^2 - 10xy + 3y^2$

SECTION 8.6
SOLVING QUADRATIC EQUATIONS
BY FACTORING

A quadratic equation is a second degree equation. These equations can have at the most two solutions. The degree of the equation tells how many possible solutions the equation can have. We saw in chapter 3 that one degree equations had at the most one solution.

We will use common factoring, difference of squares, trinomials part 1, and trinomials part 2 to solve quadratic equations. Chapter 10 will have another method that can solve all types of quadratic equations.

The key to solving any quadratic equation is that the equation must be set to zero. Any equation that is not set to zero must be set to zero using the techniques from chapter 3. All examples will show either an equation set to zero or not.

EXAMPLE 1:

Solve for h: $h^2 + h = 0$

Determine the common factor and then factor it out.
$h(h + 1) = 0$

Set each factor to zero.
$h = 0$ $h + 1 = 0$

Solve both equation using methods from chapter 3
$h = 0$ $h + 1 = 0$
$$\underline{\quad -1 \quad -1}$$
$$h \quad = -1$$

The solutions are 0 and -1. Notice one factor was set to zero, and did not need to be solved.

Check: $h = 0$ $h = -1$
$h(h + 1) = 0$ $h(h + 1) = 0$
$0(0 + 1) = 0$ $-1(-1 + 1) = 0$
$0 \cdot 1 = 0$ $-1 \cdot 0 = 0$
$0 = 0$ $0 = 0$ both check

0 and -1 are indeed the solutions. ■

EXAMPLE 2:

Solve for n: $n^2 = 16$
Set equation to zero.

$n^2 = 16$
$\underline{-16 \quad -16}$
$n^2 - 16 = 0$

Factor the difference of squares.
$(n - 4)(n + 4) = 0$

Set each factor to zero.
$n - 4 = 0 \qquad n + 4 = 0$

Solve both equation using methods from chapter 3
$n - 4 = 0 \qquad n + 4 = 0$
$\underline{+4 \;\; +4} \qquad \underline{-4 \;\; -4}$
$n = 4 \qquad\qquad n = -4$

The solutions are 4 and -4.

Check: $n = 4 \qquad\qquad n = -4$
$n^2 = 16 \qquad\qquad n^2 = 16$
$4^2 = 16 \qquad\qquad (-4)^2 = 16$
$16 = 16 \qquad\qquad 16 = 16 \quad$ both check

Parentheses were placed around -4 because the entire number was used for the substitution of n. ∎

EXAMPLE 3:

Solve for u: $u^2 - 6u - 7 = 0$

Factor the trinomial.
$(u - 7)(u + 1) = 0$

Set each factor to zero.
$u - 7 = 0 \qquad u + 1 = 0$

Solve both equation using methods from chapter 3
$u - 7 = 0 \qquad u + 1 = 0$
$\underline{+7 \;\; +7} \qquad \underline{-1 \;\; -1}$
$u = 7 \qquad\qquad u = -1$

7 and -1 are the solutions.

Check: $u = 7$ $u = -1$

$u^2 - 6u - 7 = 0$ $u^2 - 6u - 7 = 0$

$7^2 - 6 \cdot 7 - 7 = 0$ $(-1)^2 - 6(-1) - 7 = 0$

$49 - 6 \cdot 7 - 7 = 0$ $1 - 6(-1) - 7 = 0$

$49 - 42 - 7 = 0$ $1 + 6 - 7 = 0$

$7 - 7 = 0$ $7 - 7 = 0$

$0 = 0$ $0 = 0$ both check.

Notice that the order of operations was used in the check. ■

EXAMPLE 4:

Solve for j: $8j^2 + 6j = 5$

Set equation to zero.

$8j^2 + 6j = 5$

$\underline{ -5 \quad -5}$

$8j^2 + 6j - 5 = 0$

Factor using the grouping method

$8j^2 + 6j - 5 = 0$

First multiply the first coefficient and the third term.

$8(-5) = -40$

Rewrite -40 as a difference of -6 with j

$8j^2 - 4j + 10j - 5 = 0$

Place parentheses around the first two terms and the last two terms and then factor out common factors from each group of terms.

$(8j^2 - 4j) + (10j - 5) = 0$

$4j(2j - 1) + 5(2j - 1) = 0$

Factored equation is: $(2j - 1)(4j + 5) = 0$

Set each factor to zero.

$2j - 1 = 0$ $4j + 5 = 0$

Solve both equation using methods from chapter 3

$2j - 1 = 0$ $4j + 5 = 0$

$\underline{ +1 \quad +1}$ $\underline{ -5 \quad -5}$

$2j = 1$ $4j = -5$

$$\frac{2j}{2} = \frac{1}{2} \qquad \frac{4j}{4} = \frac{-5}{4}$$

$$j = 1/2 \qquad j = -5/4$$

$1/2$ and $-5/4$ are the solutions.

Check: $j = 1/2 \qquad\qquad j = -5/4$

$8j^2 + 6j = 5 \qquad\qquad\quad 8j^2 + 6j = 5$

$8(1/2)^2 + 6(1/2) = 5 \qquad 8(-5/4)^2 + 6(-5/4) = 5$

$8(1/4) + 6(1/2) = 5 \qquad 8(25/16) + 6(-5/4) = 5$

$2 + 3 = 5 \qquad\qquad\qquad 25/2 \quad - 15/2 = 5$

$5 = 5 \qquad\qquad\qquad\qquad 10/2 = 5$

$\qquad\qquad\qquad\qquad\qquad\quad 5 = 5 \quad$ both check. ∎

Many students feel that if fractions come up as solutions that there is a mistake in the problem. Example 4 shows that fractions are perfectly legitimate solutions.

The first four examples all had 2 solutions. Now let us look at an example that has 1 solution and at two examples that have no solutions.

EXAMPLE 5:

Solve for x: $\quad x^2 + 10x + 25 = 0$

Factor the trinomial.
$(x + 5)(x + 5) = 0$

Set each factor to zero.
$x + 5 = 0 \qquad x + 5 = 0$

Solve both equation using methods from chapter 3
$$x + 5 = 0 \qquad\quad x + 5 = 0$$
$$\underline{\quad -5 \quad -5} \qquad \underline{\quad -5 \quad -5}$$
$$x = -5 \qquad\qquad x = -5$$

-5 is the only solution since both factors were solved to -5

Check: $x = -5$
$x^2 + 10x + 25 = 0$
$(-5)^2 + 10(-5) + 25 = 0$
$25 - 50 + 25 = 0$
$-25 + 25 = 0$
$0 = 0 \quad$ checks. ∎

EXAMPLE 6:

Solve for x: $x^2 - x = -1$

Set equation to zero.
$$x^2 - x = -1$$
$$\underline{ + 1 \quad + 1}$$
$$x^2 - x + 1 = 0$$

Factor the trinomial.
$x^2 - x + 1$ cannot be factored.

$x^2 - x = -1$ has no solutions. ■

EXAMPLE 7:

Solve for y: $y^2 + 9 = 0$

$y^2 + 9$ cannot be factored.

There are no solutions for $y^2 + 9 = 0$. ■

Examples 6 and 7 have no solutions since they have expressions that cannot be factored. These two examples will be used again in Chapter 10 and an explanation on why there is no solution will be given in chapter 10.

Now let us see what happens when there is a numerical common factor involved.

EXAMPLE 8:

Solve: $9m^2 + 18m - 27 = 0$

Factor the trinomial.
$$9(m^2 + 2m - 3) = 0$$
$$9(m - 1)(m + 3) = 0$$

Set each factor to zero.
$$9 = 0 \qquad m - 1 = 0 \qquad m + 3 = 0$$

Notice that $9 = 0$ is a false statement and cannot be solved. This false statement can be ignored. Solve the other two statements.
$$m - 1 = 0 \qquad\qquad m + 3 = 0$$
$$\underline{ + 1 \quad + 1} \qquad\qquad \underline{ - 3 \quad - 3}$$
$$m = 1 \qquad\qquad\qquad m = -3$$

Check: $m = 1$ $m = -3$

$9m^2 + 18m - 27 = 0$ $9m^2 + 18m - 27 = 0$

$9 \cdot 1^2 + 18 \cdot 1 - 27 = 0$ $9(-3)^2 + 18(-3) - 27 = 0$

$9 \cdot 1 + 18 \cdot 1 - 27 \cdot 1 = 0$ $9 \cdot 9 + 18(-3) - 27 = 0$

$9 + 18 - 27 = 0$ $81 - 54 - 27 = 0$

$27 - 27 = 0$ $27 - 27 = 0$

$0 = 0$ $0 = 0$ both check. ∎

SUMMARY: Set equations to zero if necessary

Factor left side of equal sign

Solve factors

PROBLEMS:

Solve for the given variable.

1) $m^2 - 2m - 3 = 0$ 2) $n^2 - 7n + 6 = 0$

3) $o^2 + 8o + 15 = 0$ 4) $p^2 + 4p - 21 = 0$

5) $q^2 - 4q = 12$ 6) $r^2 + 5r = 14$

7) $2s^2 + 5s - 3 = 0$ 8) $4t^2 - 24t + 35 = 0$

9) $4u^2 + 11u = -6$ 10) $6s^2 + 19s + 19 = 0$

11) $y^2 - 14y + 49 = 0$ 12) $z^2 + 5z + 4 = 0$

13) $a^2 + 3a - 10 = 0$ 14) $b^2 - 3b - 18 = 0$

15) $c^2 - 12c + 32 = 0$ 16) $d^2 + 8d = -15$

17) $h^2 = 11h - 24$ 18) $3g^2 + 7g + 2 = 0$

19) $6h^2 + 11h - 10 = 0$ 20) $5j^2 + 2j = 3$

21) $16p^2 + 40p + 25 = 0$ 22) $o^2 - 12o - 35 = 0$

23) $5q^2 + 13q = 6$ 24) $r^2 - 2r = 0$

25) $s^2 = -8s$ 26) $5t^2 - 15t = 0$

27) $u^2 - 25 = 0$

28) $v^2 = 81$

29) $2w^2 - 18 = 0$

30) $3y^2 + 24y + 45 = 0$

31) $6z^2 + 28z = 10$

32) $9y^2 - 12y + 4 = 0$

33) $4h^2 = 13h + 12$

34) $j^2 + 5j = 0$

35) $k^2 = 7k$

36) $4m^2 + 20m = 0$

37) $n^2 = 49$

38) $o^2 = 64$

39) $3p^2 - 75 = 0$

40) $4q^2 - 4q = 24$

41) $15r^2 + 27r = 6$

42) $g^2 + 8g + 16 = 0$

43) $v^2 + 10v + 25 = 0$

44) $j^2 + j = 2$

45) $k^2 - 3k = 7$

46) $m^2 + 4m + 5 = 0$

47) $16n^2 - 64 = 0$

48) $9o^2 - 36 = 0$

49) $16u^2 - 42u + 5 = 0$

50) $u^2 = -2u + 35$

SECTION 8.7
LITERAL EQUATIONS
SOLUTIONS BY FACTORING

Solving literal equations is a two step process. The first step is to factor out the unknown that is needed, and then divide out what is not needed. The only restriction is that the needed unknown is on one side of the equal sign.

EXAMPLE 1:

Solve for L: $aL + bL = c$

Factor out common L
$L(a + b) = c$

Divide by binomial
$$\frac{L(a + b)}{a + b} = \frac{c}{a + b}$$

The solution is: $L = \dfrac{c}{a + b}$

Check: $L = \dfrac{c}{a + b}$

$aL + bL = c$

$$a\left(\frac{c}{a + b}\right) + b\left(\frac{c}{a + b}\right) = c$$

Multiply through by $(a + b)$

$$a\left(\frac{c}{a + b}\right)(a + b) + b\left(\frac{c}{a + b}\right)(a + b) = c(a + b)$$

$ac + bc = c\,(a + b)$
$ac + bc = ca + cb$
$ac + bc = ac + bc$ ■

The multiplication to clear fractions will be explained in Chapter 9. Notice that the commutative property was used in the last step of the check to make both sides match.

EXAMPLE 2:

Solve for f: $\quad fp = fq + g$

Get f to one side of the equal sign by using methods from Chapter 3

$fp = fq + g$
$\underline{-fq \quad -fq}$
$fp - fq = g$

Factor out common f
$f(p - q) = g$

Divide by binomial
$$\frac{f(p-q)}{p-q} = \frac{g}{p-q}$$

The solution is: $f = \dfrac{g}{p-q}$

Check: $f = \dfrac{g}{p-q}$

$fp = fq + g$

$$\left(\frac{g}{p-q}\right)p = \left(\frac{g}{p-q}\right)q + g$$

Multiply through by $(p - q)$

$$\left(\frac{g}{p-q}\right)(p-q)p = \left(\frac{g}{p-q}\right)(p-q)q + g(p-q)$$

$gp = gq + g(p-q)$
$gp = gq + gp - gq$
$gp = gp$ ■

Notice that the distribution was done from the back of the parentheses. The distributive property works the same as if the multiplication took place in front of the parentheses.

Now let us look at an example with a distribution in it.

EXAMPLE 3:

Solve for k: $x(j-k)=jk$

Use distribution first
$jx - kx = jk$

Get k to one side of the equal sign my using methods from Chapter 3
$jx - kx = jk$
$\underline{\quad + kx \quad + kx\quad}$
$jx = jk + kx$

Factor out common k
$jx = k(j+x)$

Divide by binomial
$$\frac{jx}{j+x} = \frac{k(j+x)}{j+x}$$

The solution is: $\dfrac{jx}{j+x} = k$

Check: $\dfrac{jx}{j+x} = k$

$x(j-k)=jk$

$$x\left(j - \frac{jx}{j+x}\right) = j\left(\frac{jx}{j+x}\right)$$

Do multiplication on both sides.
$$\left(jx - \frac{jx^2}{j+x}\right) = \left(\frac{j^2x}{j+x}\right)$$

Multiply through by $(j+x)$
$$(j+x)\left(jx - \frac{jx^2}{j+x}\right) = \left(\frac{j^2x}{j+x}\right)(j+x)$$
$(j+x)jx - jx^2 = j^2x$
$j^2x + jx^2 - jx^2 = j^2x$
$j^2x = j^2x$ ∎

It does not matter on what side of the equal sign is the needed unknown with literal equations. The procedure for the check will be explained in Chapter 9.

SUMMARY: Get needed unknown to one side of the equation.

Factor out needed unknown

Divide out what is not needed

PROBLEMS:

1) Solve for h: $dh + gh = f$

2) Solve for j: $fj = gj + h$

3) Solve for k: $n(j - k) = jk$

4) Solve for L: $C = L - sL$

5) Solve for j: $n(j - k) = jk$

6) Solve for y: $my - ny = p$

7) Solve for q: $pq = p - 2q$

8) Solve for t: $s(t + v) = tv$

9) Solve for v: $s(t + v) = tv$

10) Solve for L: $C = L + sL$

11) Solve for p: $fp + fq = pq$

12) Solve for q: $fp + fq = pq$

SECTION 8.8
SOLVING WORD PROBLEMS BY FACTORING

The translation of word problems is exactly the same. There are no new words that need to be translated. The only difference is that there may be two solutions. The key to solving word problems in this section is to be careful of what the problem is asking you to find. A problem may have fractions in the solution, but if the problem asks for integers then there is no solution to the problem. A problem may have positive and negative solutions, but if the problem only asks for positive solutions then the negative solution is not needed. Each example will have an explanation on the nature of the solution.

EXAMPLE 1:

One number is 6 more than another. If their product is 27, find the two integers.

Translate: $x(x + 6) = 27$
$x(x + 6)$ represents the product of the two numbers.

Multiply: $x(x + 6) = 27$
$x^2 + 6x = 27$

Set the equation to 0: $x^2 + 6x = 27$
$$\underline{\quad -27 \quad\quad -27 \quad}$$
$x^2 + 6x - 27 = 0$

Factor trinomial: $x^2 + 6x - 27$
$(x + 9)(x - 3)$

Set factors to zero: $x + 9 = 0 \qquad x - 3 = 0$

Solve: $x + 9 = 0 \qquad x - 3 = 0$
$$\underline{\quad -9 \quad -9 \quad} \quad \underline{\quad +3 \quad +3 \quad}$$
$x = -9 \qquad x = 3$ This part is the number. Now find 6 more than

$x + 6 = -3 \qquad x + 6 = 9$ 6 was added to both numbers.

Check: $-9(-3) = 27 \quad 3 \cdot 9 = 27$
$\quad\quad\quad 27 = 27 \quad\quad\quad 27 = 27$ both check

The solutions are −3 and −9, or 3 and 9.

A formal check is not needed. Just check the product of the two numbers. ■

It may seem like that Example 1 has four solutions, but there are only two. There are two sets of solutions. Recall that the product of two negative numbers equals a positive number as well as the product of two positive numbers. Since there are two sets of answers, the number of possible solutions for a quadratic equation still holds.

EXAMPLE 2:

The sum of an integer and its square is 24. Find the integer

Translate: $x + x^2 = 42$
$x + x^2$ represents the sum of the two numbers.

Rewrite: $x + x^2 = 42$
$x^2 + x = 42$ Commutative property was used to put equation in standard form.

Set the equation to 0: $x^2 + x = 42$
$$\underline{-42 \quad -42}$$
$x^2 + x - 42 = 0$

Factor trinomial: $x^2 + x - 42$
$(x + 7)(x - 6)$

Set factors to zero: $x + 7 = 0 \qquad x - 6 = 0$

Solve: $x + 7 = 0 \qquad x - 6 = 0$
$$\underline{-7 \quad -7} \qquad \underline{+6 \quad +6}$$
$x = -7 \qquad x = 6$

Check: $x = -7$ $\qquad x = 6$
$x + x^2 = 42$ $\qquad x + x^2 = 42$
$-7 + (-7)^2 = 42$ $\qquad 6 + 6^2 = 42$
$-7 + 49 = 42$ $\qquad 6 + 36 = 42$
$42 = 42$ $\qquad 42 = 42$ Both check

The solution is −7 or 6.

A formal check was needed because it was a sum of an integer and its square. ∎

EXAMPLE 3:

One integer is 9 less than twice another. If the product of the integers is 35, what are the two integers?

Translate: $x(2x - 9) = 35$
$x(2x - 9)$ represents the product of the two numbers.

Multiply: $x(2x - 9) = 35$
                $2x^2 - 9x = 35$

Set the equation to 0: $2x^2 - 9x = 35$
$$\frac{-35 \quad\quad -35}{2x^2 - 9x - 35 = 0}$$

Factor trinomial: $2x^2 - 9x - 35$ Use grouping method

First multiply the first coefficient and the third term.
$2(-35) = -70$

Rewrite -70 as a difference of -9 with x
$2x^2 - 14x + 5x - 35$

Place parentheses around the first two terms and the last two terms and then factor out common factors from each group of terms.
$(2x^2 - 14x) + (5x - 35)$

$2x(x - 7) + 5(x - 7)$

$(x - 7)(2x + 5)$ is the factored form.

Set factors to zero: $x - 7 = 0$ $2x + 5 = 0$

Solve: $x - 7 = 0$ $2x + 5 = 0$
$$\frac{+7 \quad +7}{x = 7}$$ $$\frac{-5 \quad -5}{2x = -5}$$
$$\frac{2x}{2} = \frac{-5}{2}$$
$$x = \frac{-5}{2}$$

Only use 7 because the problem is asking for integers.
This part is the number. Now find 9 less than twice another.

$2x - 9 = 2 \cdot 7 - 9 = 14 - 9 = 5$ 7 was substituted

Check: $5 \cdot 7 = 35$ check

The solution set is 5 and 7

A formal check is not needed. Just check the product of the two numbers. ■

Only one set of solutions occurred. The problem asked for integers. The fractional solution was thrown out because it was not needed.

EXAMPLE 4:

A rectangle is 2cm longer than it is wide. Find the dimensions of the rectangle if the area of the rectangle is 80 sq. cm.

First draw a picture of the rectangle with the length and width labeled.

x + 2

x

Turn labels into an equation. Use $LW = A$. (area of a rectangle.)
$x(x + 2) = 80$

Multiply: $x(x + 2) = 80$
$x^2 + 2x = 80$

Set equation to 0: $x^2 + 2x = 80$
$\underline{-80 - 80}$
$x^2 + 2x - 80 = 0$

Factor trinomial: $x^2 + 2x - 80$
$(x - 8)(x + 10)$

Set factors to 0: $x - 8cm = 0$ $x + 10cm = 0$

Solve equations: $x - 8cm = 0$ $x + 10cm = 0$
$\underline{+ 8cm \quad + 8cm}$ $\underline{- 10cm \quad - 10cm}$
$x = 8cm$ $x = -10cm$

Use $x = 8cm$ to finish problem. $x = -10cm$ represents a negative width. There is no such thing as a negative width nor a negative length. Measurements are never negative.
$x = 8cm$ represents the width.

$x + 2$cm $= 10$cm represents the length.

The dimensions of the rectangle are 8cm and 10cm.

Check: 8cm $\cdot$ 10cm $= 80$ sq. cm verified. ■

The cm was placed in the equations to show the unit of measure.

SUMMARY: Translate problem

Do any operation that is necessary

Set equation to 0

Factor polynomials

Set factors to 0

Solve equations.

Throw out any solutions that are not needed.

Solve problem.

PROBLEMS:

1)   A square integer minus 3 times the same integer is 4. Find the integer.

2)   A square integer plus 5 times the same integer is 14. Find the integer.

3)   One integer is 5 less than another. If the product of the two integers is 66, what are the integers?

4)   One integer is 7 more than another. If the product of the two integers is 60, what are the integers?

5)   One integer is 1 less than twice another. If the product of the two integers is 120, what are the integers?

6)   One integer is 2 more than 3 times another. If the product of the two integers is 56, what are the integers?

7)   The sum of an integer and its square is 72. What is the integer?

8)	The square of an integer is 56 more than the integer. Find the integer.

9)	The length of a rectangle is 8 feet longer than its width. What are the dimensions of the rectangle if the area is 65 square feet?

10)	The length of a rectangle is 4 centimeters longer than its width. What are the dimensions of the rectangle if the area is 140 square centimeters?

11)	The width of a rectangle is 3 feet less than its length. What are the dimensions of the rectangle if the area is 70 square feet?

12)	The length of a rectangle is 5 centimeters longer than its width. What are the dimensions of the rectangle if the area is 150 square centimeters?

13)	One integer is 4 less than another. If the product of the two integers is 117, what are the integers?

14)	One integer is 9 more than another. If the product of the two integers is 52, what are the integers?

15)	One integer is 1 less than 3 times another. If the product of the two integers is 30, what are the integers?

16)	One integer is 4 less than 3 times another. If the product of the two integers is 55, what are the integers?

Refer back to section 3.5

17)	The product of two consecutive integers is 132. Find the two integers.

18)	Find two consecutive positive even integers if their product is 120.

CHAPTER 9
ALGEBRAIC FRACTIONS

SECTION 9.1
AN INTRODUCTION TO ALGEBRAIC FRACTIONS

An algebraic fraction is a fraction with at least one unknown in it. The unknowns can appear in the numerator, and in the denominator. If an unknown appears in the numerator then there is no problem. If an unknown appears in the denominator then a problem may exist. Examples 2 and 3 will show what happens when there are unknowns in the numerator and denominator. Example 1 only shows what algebraic fractions look like.

EXAMPLE 1:

a) $\dfrac{x}{y}$ b) $\dfrac{z}{z + 5}$ c) $\dfrac{a^2 + 4a + 3}{a + 1}$

All are examples of algebraic fractions. ▪

EXAMPLE 2:

$$\dfrac{b}{6}$$

Lets substitute a number in for b. Let be = 3.

$$\dfrac{b}{6} = \dfrac{3}{6} = \dfrac{1}{2}$$

A fraction still occurs when $b = 3$. ▪

As long as an unknown is in the numerator, there will always be a fraction whenever a number is placed into the unknown.

EXAMPLE 3:

$$\dfrac{3}{c}$$

Let's substitute a number in for c. Let $c = 5$.

$$\dfrac{3}{5}$$

A fraction still occurs when $c = 5$.
Now let $c = 0$.

$$\frac{c}{0} \blacksquare$$

Division by 0 cannot happen. Zero is not acceptable in the denominator. Zero is never allowed in the denominator when there is a single unknown in the denominator. Now let us see what happens when there is more than 1 term in the denominator.

EXAMPLE 4:

a) $$\frac{1}{d+2}$$

Example 3 showed that zero cannot be in the denominator. If we substitute 0 in for d, we get 2 in the denominator. It sounds like 0 can be used when there is more than 1 term involved in the denominator. 0 can be used in the denominator when the calculation provides another answer other than 0. The question really is when does the denominator go to zero. Example 3 showed that when there is a single term in the denominator then the denominator goes to zero at zero. Example 4a is an example of a denominator that goes to zero at another number. To find the number that sends the denominator to 0, just set the denominator to zero and solve.

$$d + 2 = 0$$
$$\underline{-2 \quad -2}$$
$$d = -2$$

-2 is the number that sends the denominator to 0.

b) $$\frac{3g}{g^2 - 2g + 3}$$

Find the number or numbers that send the denominator to 0.

Set the trinomial to zero: $\qquad g^2 - 2g + 3 = 0$

Factor the trinomial: $\quad g^2 - 2g + 3$
$$(g-2)(g-1)$$

Set factors to 0: $\qquad g - 2 = 0 \qquad g - 1 = 0$

Solve equations: $\qquad g - 2 = 0 \qquad g - 1 = 0$
$$\underline{+2 \quad +2} \qquad \underline{+1 \quad +1}$$
$$g = 2 \qquad\qquad g = 1$$

2 and 1 are the numbers that send the denominator to zero. It is left to the student to check the solutions in the denominator. At this stage of the text, students should be able to check the solutions without any guidance. ■

Notice that the numerator was not used. Any number is allowed in the numerator. We are only interested in when the denominator goes to zero.

<u>The numbers that send the denominators to zero are known as excluded numbers.</u> These numbers are not allowed when solving equations. Solving equations will be studied in Section 9.7.

SUMMARY: Set denominator to zero
Solve equation.

PROBLEMS:

Are the following Algebraic Fractions?

1) $\dfrac{a}{10}$

2) $\dfrac{1}{18}$

3) $\dfrac{19}{b+7}$

4) $c - 10$

5) $\dfrac{10s^8 + 5s^6 - 10s^4}{5s^3}$

6) $7t$

Find the excluded value or values. Write none if there are no excluded values.

7) $\dfrac{17}{u}$

8) $\dfrac{v-5}{2}$

9) $\dfrac{7}{w^2 - 9}$

10) $\dfrac{y+3}{y^2 - 7y + 12}$

11) $\dfrac{2z-1}{3z^2 + z - 2}$

12) $\dfrac{a}{8}$

13) $\dfrac{4}{b+3}$

14) $\dfrac{c-1}{c-5}$

SECTION 9.2
REDUCING ALGEBRAIC FRACTIONS

We saw a reduction of algebraic fractions back in Section 6.4. We will only concentrate on fractions that have factorable numerators and denominators in this section. Polynomials can be eliminated in the same way as numerical reduction of fraction problems. Polynomials can be eliminated vertically or diagonally. Polynomials must match exactly to be eliminated. The remaining result is the reduced fraction once the elimination is completed.

EXAMPLE 1:

Reduce each fraction.

a) $\dfrac{12}{40} = \dfrac{2*2*3}{2*2*2*5} = \dfrac{3}{2*5} = \dfrac{3}{10}$

Factoring was done first. Matching 2's were eliminated. The remaining 2 in the denominator could not be eliminated. There were three 2's in the denominator, but only two 2's in the numerator. Only exact matches can be eliminated.

b) $\dfrac{-8}{54} = \dfrac{-1*2*2*2}{2*3*3*3} = \dfrac{-1*2*2}{3*3*3} = \dfrac{-4}{27}$

Factoring was done first. Matching 2s were eliminated. The remaining 2s in the numerator could not be eliminated. There were three 2s in the numerator, but only one 2 in the denominator. Only exact matches can be eliminated. The 3s and the negative 1 could not be eliminated.

Recall that fractions represent division. A negative number divided by a positive number results in a negative number. That is why the result in b is negative.

c) $\dfrac{6r-12}{r^2-4} = \dfrac{6(r-2)}{(r+2)(r-2)} = \dfrac{6}{r+2}$

Factoring was done first. $(r-2)$ was eliminated.

d) $\dfrac{8d^2-8}{d^2-8d-9} = \dfrac{8(d^2-1)}{(d+1)(d-9)} = \dfrac{8(d+1)(d-1)}{(d+1)(d-9)} = \dfrac{8(d-1)}{d-9}$

Factoring was done first. Notice that the numerator was factored twice. $(d+1)$ was eliminated.

e) $\quad \dfrac{2h^2 + h - 1}{2h^2 - h - 3} = \dfrac{(2h-1)(h+1)}{(2h-3)(h+1)} = \dfrac{2h-1}{2h-3}$

Factoring was done first. $(h + 1)$ was eliminated. ■

EXAMPLE 2:

Reduce each fraction.

a) $\quad \dfrac{6h^2 + 3h}{9h^2 + 15h} = \dfrac{3h(2h+1)}{3h(3h+5)} = \dfrac{2h+1}{3h+5}$

Factoring was done first. $3e$ was eliminated.

b) $\quad \dfrac{g^2 - 4}{g^2 - g - 6} = \dfrac{(g+2)(g-2)}{(g+2)(g-3)} = \dfrac{g-2}{g-3}$

Factoring was done first. $(g + 2)$ was eliminated. ■

The next example originally appeared in Example 4 in Section 6.4. Recall that a fraction can represent division.

EXAMPLE 3:

$(j^2 + 7j + 12) \div (j + 4)$ original form

Rewrite problem as a fraction instead of doing long division.

$$\dfrac{j^2 + 7j + 12}{j + 4} = \dfrac{(j+3)(j+4)}{j+4} = j + 3$$

Factoring was done first. $(j + 4)$ was eliminated. ■

CAUTION: This method only works when the division problem does not have a remainder.

Sometimes a -1 occurs when an Algebraic fraction is reduced. Algebraic fractions of the form $\dfrac{a-b}{b-a} = -1$. I will show why this occurs using long division.

First set up $\dfrac{a-b}{b-a}$ as a long division problem.

$b-a \overline{)-b+a}$ The commutative property was used so that the division is more clearer.

$$b-a \overline{)-b+a}^{-1} \quad (-b)/b = -1.\text{ Refer back to section 6.4 if this division is not clear.}$$

$$\underline{-b+a}$$
$$0$$

EXAMPLE 4:

Reduce the fraction.

$$\frac{2x-6}{9-x^2} = \frac{2(x-3)}{(3+x)(3-x)} = \frac{-2}{3+x}$$

$\dfrac{x-3}{3-x}$ is of the form $\dfrac{a-b}{b-a}$. $\dfrac{x-3}{3-x} = -1$. $-1 * 2 = -2$. ■

PROBLEMS:

Reduce the following fractions.

1) $\dfrac{9}{30}$

2) $\dfrac{3}{75}$

3) $\dfrac{-7}{77}$

4) $\dfrac{-4}{54}$

5) $\dfrac{3j+18}{5j+30}$

6) $\dfrac{9k+12}{21k+28}$

7) $\dfrac{5m-5}{m^2+m-2}$

8) $\dfrac{5d^2-20}{d^2+4d-12}$

9) $\dfrac{2h^2+3h-5}{2h^2+11h+15}$

10) $\dfrac{6g^2-g-2}{3g^2-5g+12}$

11) $\dfrac{r^2 - 12r + 35}{r^2 - 9r + 14}$

12) $\dfrac{q^2 - 2q - 3}{q^2 + 5q - 24}$

13) $\dfrac{12h^2 + 4h}{16h^2 + 20h}$

14) $\dfrac{14g^2 + 91g}{21g^2 + 42g}$

15) $\dfrac{h^2 - 6h - 16}{h^2 - 64}$

16) $\dfrac{j^2 - 25}{j^2 - j - 20}$

17) $\dfrac{2k - 10}{25 - k^2}$

18) $\dfrac{3m - 12}{16 - m^2}$

19) $\dfrac{n - 11}{11 - n}$

20) $\dfrac{o - 29}{29 - o}$

21) $\dfrac{p + 3}{p^2 - 9}$

22) $\dfrac{q^2 - 9}{3 - q}$

23) $\dfrac{r^2 + 10r + 25}{r + 5}$

24) $\dfrac{s + 8}{s^2 + 2s - 48}$

25) $\dfrac{t^2 - 7t + 6}{t^2 - 2t + 1}$

26) $\dfrac{2u^2 - 13u + 15}{2u^2 + 5u - 12}$

Can the following division problems be solved using the methods from this section? If so, work the problem and write out the answer.

27) $(b^2 + 9b + 14) \div (b + 7)$

28) $(g^2 + 2g - 35) \div (g - 5)$

29) $(t^2 - 7t + 10) \div (t - 2)$

30) $(4d^2 - 18d - 15) \div (d - 5)$

SECTION 9.3
MULTIPLYING & DIVIDING ALGEBRAIC FRACTIONS

Multiplication of algebraic fractions is similar to the multiplication of numerical fractions, Recall that cancellation could occur when multiplying numerical fractions. I will show multiplication of numerical fractions first to remind readers how to multiply fractions.

EXAMPLE 1:

$$\frac{8}{45}*\frac{9}{44}=\frac{2*2*2}{5*3*3}*\frac{3*3}{2*2*11}=\frac{2}{5}*\frac{1}{11}=\frac{2}{55}\ \blacksquare$$

The only operation to do is to multiply across the numerator and the denominator after factoring out the numerator and the denominator once cancellations are done. The answer is in reduced form.

EXAMPLE 2:

a) $\dfrac{h}{8}*\dfrac{5}{2h^2}=\dfrac{h}{2*2*2}*\dfrac{5}{2hh}=\dfrac{5}{16h}$

b) $\dfrac{j^2-64}{j^2+10j+25}*\dfrac{j^2-4j-45}{j^2-j-56}=\dfrac{(j+8)(j-8)}{(j+5)(j+5)}*\dfrac{(j+5)(j-9)}{(j-8)(j+7)}=\dfrac{(j+8)(j-9)}{(j+5)(j+7)}$

c) $\dfrac{6k^2-48k-120}{k^2+8k-33}*\dfrac{k^2+6k-55}{k^2-6k-40}=\dfrac{6(k+2)(k-10)}{(k-3)(k+11)}*\dfrac{(k-5)(k+11)}{(k-10)(k+4)}=\dfrac{6(k+2)(k-5)}{(k-3)(k+4)}\ \blacksquare$

Notice that the answers for b and c are left in factored form. Some instructors may want the answers multiplied. Be careful when multiplying answers. Instructors have been known to take off points when students incorrectly multiply the answers. In my opinion, an answer left in factored form is already in reduced form.

Division of algebraic fractions is similar to the division of numerical fractions, Recall that the fraction on the right side of the division symbol gets inverted. Inverted is another way of saying flip over or turn upside down. Problems then take the form of multiplication problems. The division rule of fractions is that the problem becomes a multiplication problem once the inversion has occurred. I call division of fractions pan cake problems. Pan cakes need to be flipped over when cooked. Fractions need to be flipped over before they can be multiplied. I will show division of numerical fractions first to remind readers how to divide fractions.

EXAMPLE 3:

$$\frac{9}{38} \div \frac{3}{44} = \frac{9}{38} * \frac{44}{3} = \frac{3*\cancel{3}}{\cancel{2}*19} * \frac{\cancel{2}*2*11}{\cancel{3}} = \frac{3}{19} * \frac{22}{1} = \frac{66}{19} \; \blacksquare$$

Answers can remain as an improper fraction. <u>An improper fraction is when the numerator is larger than the denominator.</u>

EXAMPLE 4:

a)
$$\frac{p^3 - p^2 + p}{p} \div \frac{3p^2 - 6p + 9}{12}$$

$$= \frac{p^3 - p^2 + p}{p} * \frac{12}{3p^2 - 6p + 9}$$

$$= \frac{p(p^2 - p + 1)}{p} * \frac{12}{3(p^2 - 2p + 3)} = \frac{4(p^2 - p + 1)}{p^2 - 2p + 3}$$

b)
$$\frac{49m^2 - 100n^2}{7m^2 + 11mn - 30n^2} \div \frac{7m^2 + 59mn + 70n^2}{m^2 + 2mn - 3n^2}$$

$$= \frac{49m^2 - 100n^2}{7m^2 + 11mn - 30n^2} * \frac{m^2 + 2mn - 3n^2}{7m^2 + 59mn + 70n^2}$$

$$= \frac{(7m + 10n)(7m - 10n)}{(7m - 10n)(m + 3n)} * \frac{(m - n)(m + 3n)}{(7m + 10n)(m + 7n)} = \frac{m - n}{m + 7n}$$

c)
$$\frac{o^2 + 10o + 21}{20o^2 - 33o + 10} \div \frac{5o^2 + 3o - 2}{o^2 + 7o + 12}$$

$$= \frac{o^2 + 10o + 21}{20o^2 - 33o + 10} * \frac{5o^2 + 3o - 2}{o^2 + 7o + 12} = \qquad \blacksquare$$

$$\frac{(o + 7)(o + 3)}{(4o - 5)(5o - 2)} * \frac{(5o - 2)(o + 1)}{(o + 3)(o + 4)} = \frac{(o + 7)(o + 1)}{(4o - 5)(o + 4)}$$

None of the factoring rules from the chapter 8 have changed. Readers should refer back to chapter 8 in case there are difficulties in factoring out the problems in this section.

PROBLEMS:

Do the indicated operation, and write answers in reduced form where possible.

1) $\dfrac{1}{10} * \dfrac{4}{13}$

2) $\dfrac{3}{44} * \dfrac{11}{63}$

3) $\dfrac{5}{12} * \dfrac{18}{45}$

4) $\dfrac{7}{50} * \dfrac{10}{14}$

5) $\dfrac{3q}{2} * \dfrac{4}{q^2}$

6) $\dfrac{p^2 + 10p}{3p^2} * \dfrac{12p}{2p + 20}$

7) $\dfrac{2u^2 + 5u - 3}{6u^2 - 19u + 10} * \dfrac{9u^2 - 12u + 4}{3u^2 + 7u - 6}$

8) $\dfrac{s^2 - 4s - 21}{3s^2} * \dfrac{s^2 + 7s}{s^2 - 49}$

9) $\dfrac{t^2 + 6t + 8}{t^2 + 8t + 15} * \dfrac{t^2 + 13t + 30}{t^2 - t - 20}$

10) $\dfrac{r^2 - 8r}{4r} * \dfrac{12r^2}{r^2 - 64}$

11) $\dfrac{5}{8} \div \dfrac{1}{4}$

12) $\dfrac{15}{16} \div \dfrac{3}{44}$

13) $\dfrac{7}{13} \div \dfrac{7}{50}$

14) $\dfrac{4}{14} \div \dfrac{1}{7}$

15) $\dfrac{8}{v^2} \div \dfrac{48}{v}$

16) $\dfrac{4w - 12}{5w + 15} \div \dfrac{8w^2}{w^2 + 3w}$

17) $\dfrac{3x^2 - 14x - 5}{6x^2 - 7x - 3} \div \dfrac{6x^2 - 25x - 25}{2x^2 - 8x - 15}$

18) $\dfrac{y^2 + 2y - 8}{9y^2} \div \dfrac{y^2 - 16}{3y - 12}$

19) $\dfrac{z^2 - 9z + 20}{z^2 - 11z + 28} \div \dfrac{z^2 - 8z + 15}{z^2 - 14z + 49}$

20) $\dfrac{6a - 18}{9a} \div \dfrac{3a - 9}{a^2 + 2a}$

SECTION 9.4
ADDING & SUBTRACTING ALGEBRAIC LIKE FRACTIONS

Like fractions are fractions that have matching denominators. All anyone has to do is add or subtract the numerators when the fractions are like. Some fractions may need to be reduced in the end. Numerical fractions will be shown to remind readers on how to add and subtract fractions.

EXAMPLE 1:

a) $\dfrac{4}{9} + \dfrac{3}{9} = \dfrac{7}{9}$

b) $\dfrac{5}{14} + \dfrac{1}{14} = \dfrac{6}{14} = \dfrac{3}{7}$ ■

It is common practice to reduce fractions to simplest form if a fraction can be reduced. Only example 1b had a reducible fraction.

EXAMPLE 2:

a) $\dfrac{9j}{14} + \dfrac{j}{14} = \dfrac{10j}{14} = \dfrac{5j}{7}$

b) $\dfrac{k+2}{k^2+7k+10} + \dfrac{k+8}{k^2+7k+10} = \dfrac{2k+10}{k^2+7k+10} = \dfrac{2(k+5)}{(k+5)(k+2)} = \dfrac{2}{k+2}$

c) $\dfrac{m+1}{m^2-4m-5} + \dfrac{m-6}{m^2-4m-5} = \dfrac{2m-5}{m^2-4m-5} = \dfrac{2m-5}{(m-5)(m+1)}$ ■

The numerator and denominator should be factored to see if the result can be reduced. Examples 2a and 2b had reducible fractions. Example 2c could not be reduced.

Combining like terms is the only computation needed when adding fractions with like denominators. The distribution rule may be needed when subtracting fractions. The distribution rule is only needed when there is more than one term in the numerator. Readers will observe the distribution rule in Example 4.

EXAMPLE 3:

a) $\dfrac{4}{10} - \dfrac{1}{10} = \dfrac{3}{10}$

b) $\quad \dfrac{6}{15} - \dfrac{12}{15} = \dfrac{-6}{15} = \dfrac{-2}{5}$ ∎

EXAMPLE 4:

a) $\quad \dfrac{9j}{14} - \dfrac{j}{14} = \dfrac{8j}{14} = \dfrac{4j}{7}$

b) $\quad \dfrac{k+2}{k^2+7k+10} - \dfrac{k+8}{k^2+7k+10} = \dfrac{k+2-(k+8)}{k^2+7k+10} = \dfrac{k+2-k-8}{k^2+7k+10} = \dfrac{-6}{(k+2)(k+5)}$

c) $\quad \dfrac{3m+4}{m^2+4m-5} - \dfrac{m-6}{m^2+4m-5} = \dfrac{3m+4-(m-6)}{m^2+4m-5} = \dfrac{3m+4-m+6}{m^2+4m-5} = \dfrac{2m+10}{m^2+4m-5}$ ∎

$\qquad = \dfrac{2(m+5)}{(m-1)(m+5)} = \dfrac{2}{m-1}$

The distribution rule was needed because there were 2 terms in the numerators in examples 4b and 4c. The entire numerator is being subtracted; therefore, the distribution rule is needed. The distribution rule is always used when there are at least 2 terms in the numerator whenever fractions are being subtracted. Students tend to forget to use the distribution rule. Some students just subtract the first term. Hopefully, my examples will help remind students that the distribution rule is needed.

Students should realize that no sign rules nor factoring rules have changed.

PROBLEMS:

Do the indicated operation, and write answers in reduced form where possible.

1) $\quad \dfrac{7}{34} + \dfrac{5}{34}$

2) $\quad \dfrac{5}{10} + \dfrac{2}{10}$

3) $\quad \dfrac{13}{36} + \dfrac{9}{36}$

4) $\quad \dfrac{5}{37} + \dfrac{11}{37}$

5) $\quad \dfrac{j}{42} + \dfrac{3j}{42}$

6) $\quad \dfrac{7k}{14} + \dfrac{3k}{14}$

7) $\quad \dfrac{7}{m+1} + \dfrac{9}{m+1}$

8) $\quad \dfrac{2n}{n+2} + \dfrac{4}{n+2}$

9) $\dfrac{o^2}{o+4} + \dfrac{3o-4}{o+4}$

10) $\dfrac{3p-1}{4} + \dfrac{p+7}{4}$

11) $\dfrac{7}{34} - \dfrac{5}{34}$

12) $\dfrac{5}{10} - \dfrac{2}{10}$

13) $\dfrac{13}{36} - \dfrac{9}{36}$

14) $\dfrac{5}{37} - \dfrac{11}{37}$

15) $\dfrac{j}{42} - \dfrac{3j}{42}$

16) $\dfrac{7k}{14} - \dfrac{3k}{14}$

17) $\dfrac{7}{m+1} - \dfrac{9}{m+1}$

18) $\dfrac{2n}{n-2} - \dfrac{4}{n-2}$

19) $\dfrac{o^2}{o-4} - \dfrac{5o-4}{o-4}$

20) $\dfrac{3p-1}{4} - \dfrac{p+7}{4}$

21) $\dfrac{4q-7}{6q} + \dfrac{2q+5}{6q}$

22) $\dfrac{4r-7}{r-5} - \dfrac{2r+3}{r-5}$

23) $\dfrac{s-7}{s^2-s-6} + \dfrac{2s-2}{s^2-s-6}$

24) $\dfrac{6t-1}{4t} - \dfrac{2t+3}{4t}$

25) $\dfrac{3u-8}{u-6} + \dfrac{u-16}{u-6}$

26) $\dfrac{5v-12}{v^2-8v+15} - \dfrac{3v-2}{v^2-8v+15}$

SECTION 9.5
ADDING & SUBTRACTING ALGEBRAIC UNLIKE FRACTIONS

Unlike fractions are fractions that have different denominators. The denominators must match before the fractions can be added or subtracted. Example 1 will show how unlike fractions can be added or subtracted together using numerical denominators.

EXAMPLE 1:

a) $\dfrac{4}{9} + \dfrac{1}{2} = \dfrac{8}{18} + \dfrac{9}{18} = \dfrac{17}{18}$

b) $\dfrac{7}{50} - \dfrac{6}{75} = \dfrac{21}{150} - \dfrac{12}{150} = \dfrac{9}{150} = \dfrac{3}{50}$

c) $\dfrac{j}{2} + \dfrac{3j}{8} = \dfrac{4j}{8} + \dfrac{3j}{8} = \dfrac{7j}{8}$

d) $\dfrac{a}{3} - \dfrac{3a}{8} = \dfrac{8a}{24} - \dfrac{9a}{24} = \dfrac{-a}{24}$ ■

It is common practice to reduce fractions to simplest form if a fraction can be reduced. Only example 1b had a reducible fraction.

Recall that a LCD(Lowest Common Denominator) was needed before the fractions could be added. Start with multiples of the highest denominator until one of the multiple is divisible by the other denominator. The next multiple of 75 is 150. 150 is divisible by 50. 150 is the LCD of 50, and 75.

Recall that the numerators must change once the denominators have changed. The new denominator is divided by the old denominator and the result is multiplied by the numerator. $\dfrac{7}{50}$ became $\dfrac{21}{150}$ by $150 \div 50 = 3$ and $3 \bullet 7 = 21$.

The LCD is needed when there are just numbers in the denominators. Now let us look at when unknowns are used in the denominators.

EXAMPLE 2:

a) $\dfrac{4}{h} + \dfrac{1}{k} = \dfrac{4k}{hk} + \dfrac{h}{hk} = \dfrac{4k + h}{hk}$

b) $\dfrac{1}{5m} - \dfrac{6}{7n} = \dfrac{7n}{35mn} - \dfrac{30m}{35mn} = \dfrac{7n - 30m}{35mn}$

c) $\dfrac{1}{2} + \dfrac{3}{p} = \dfrac{p}{2p} + \dfrac{6}{2p} = \dfrac{p+6}{2p}$

d) $\dfrac{1}{8} - \dfrac{2}{a} = \dfrac{a}{8a} - \dfrac{16}{8a} = \dfrac{a-16}{8a}$ ■

Example 2 showed that the LCD was the product of the 2 denominators. It is easier to just multiply the different denominators to make a new denominator. This will be explained further in this section.

Now let us look at examples that have binomials in the numerator.

EXAMPLE 3:

a) $\dfrac{n+4}{9} + \dfrac{n+1}{2} = \dfrac{2(n+4)}{18} + \dfrac{9(n+1)}{18} = \dfrac{2n+8}{18} + \dfrac{9n+1}{18} = \dfrac{11n+9}{18}$

b) $\dfrac{a-7}{50} - \dfrac{a-6}{75} = \dfrac{3(a-7)}{150} - \dfrac{2(a-6)}{150} = \dfrac{3a-21}{150} - \dfrac{2a-12}{150} = \dfrac{a-9}{150}$

c) $\dfrac{j+9}{2} + \dfrac{3j-1}{8} = \dfrac{4(j+9)}{8} + \dfrac{3j-1}{8} = \dfrac{4j+36}{8} + \dfrac{3j-1}{8} = \dfrac{7j-35}{8}$

d) $\dfrac{c+1}{3} - \dfrac{3c-1}{8} = \dfrac{8(c+1)}{24} - \dfrac{3(3c-1)}{24} = \dfrac{8c+8}{24} - \dfrac{9c-3}{24} = \dfrac{-c+11}{24}$ ■

Example 3 showed that the distributive rule was needed when multiplying the numerators.

Now let us look at examples that have powers in the denominators.

EXAMPLE 4:

a) $\dfrac{4}{d} + \dfrac{1}{d^2} = \dfrac{4d}{d^2} + \dfrac{1}{d^2} = \dfrac{4d+1}{d^2}$

b) $\dfrac{7}{a} - \dfrac{6}{a^4} = \dfrac{7a^3}{a^4} - \dfrac{6}{a^4} = \dfrac{7a^3-6}{a^4}$

c) $\dfrac{b+4}{b^2}+\dfrac{b-2}{b^4}=\dfrac{b^2(b+4)}{b^4}+\dfrac{b-2}{b^4}=\dfrac{b^3+4b^2}{b^4}+\dfrac{b-2}{b^4}=\dfrac{b^3+4b^2+b-2}{b^4}$

d) $\dfrac{g+1}{g^7}-\dfrac{4g-2}{g^4}=\dfrac{g+1}{g^7}-\dfrac{g^3(4g-2)}{g^7}=\dfrac{g+1}{g^7}-\dfrac{4g^4-2g^3}{g^7}=\dfrac{-4g^4+2g^3+g+1}{g^7}$ ∎

     The LCD is always the highest power when dealing with powers in the denominator. The highest power was used in the problems in example 4. Recall that the LCD of 2 and 4 would be 4. 4 is a power of 2. 4 would be the highest power of 2. $4 = 2^2$.

     All the above examples had monomials as a denominator. The next example will have monomials as well as binomials as denominators. All that is necessary to make a new denominator from binomials and monomials is simply multiply the denominators. It should be obvious that a denominator of 2 and a denominator of 3 would become a new denominator of 6. 2 and 3 are multiplied to make 6. When monomials are present with binomials in the denominator, the denominators are multiplied. It is not necessary to carry out the multiplication all the way though. Just place the two denominators together to make the new one.

EXAMPLE 5:

a) $\dfrac{8}{b}+\dfrac{2}{b+8}$ Step1: Make a new denominator by multiplying denominators together.

$\dfrac{8}{b(b+8)}+\dfrac{2}{b(b+8)}$ Step 2: Multiply the numerators by the same factor as the denominator.

$\dfrac{8(b+8)}{b(b+8)}+\dfrac{2b}{b(b+8)}$ Step 3: Use the distribution if possible.

$\dfrac{8b+64}{b(b+8)}+\dfrac{2b}{b(b+8)}$ Step 4: combine like terms.

$\dfrac{10b+64}{b(b+8)}$ Step 5: Factor out numerator if possible.

$\dfrac{2(5b+32)}{b(b+8)}$ Step 6: Cancel any factors if possible.

Nothing cancels: therefore, $\dfrac{2(5b+32)}{b(b+8)}$ is the answer in reduced form.

b) $\dfrac{1}{6} + \dfrac{1}{c+1}$ Step1: Make a new denominator by multiplying denominators together.

$\dfrac{1}{6(c+1)} + \dfrac{1}{6(c+1)}$ Step 2: Multiply the numerators by the same factor as the denominator.

$\dfrac{1(c+1)}{6(c+1)} + \dfrac{1 \bullet 6}{6(c+1)}$ Step 3: Use the distribution if possible.

$\dfrac{c+1}{6(c+1)} + \dfrac{6}{6(c+1)}$ Step 4: combine like terms.

$\dfrac{c+7}{6(c+1)}$ Step 5: Factor out numerator if possible.

$\dfrac{c+7}{6(c+1)}$ Step 6: Cancel any factors if possible.

Nothing cancels: therefore, $\dfrac{c+7}{6(c+1)}$ is the answer in reduced form.

c) $\dfrac{d}{4} - \dfrac{4}{d+6}$ Step1: Make a new denominator by multiplying denominators together.

$\dfrac{d}{4(d+6)} - \dfrac{4}{4(d+6)}$ Step 2: Multiply the numerators by the same factor as the denominator.

$\dfrac{d(d+6)}{4(d+6)} - \dfrac{4 \bullet 4}{4(d+6)}$ Step 3: Use the distribution if possible.

$\dfrac{d^2 + 6d}{4(d+6)} - \dfrac{16}{4(d+6)}$ Step 4: combine like terms.

$\dfrac{d^2 + 6d - 16}{4(d+6)}$ Step 5: Factor out numerator if possible.

$\dfrac{(d+8)(d-2)}{4(d+6)}$ Step 6: Cancel any factors if possible.

Nothing cancels: therefore, $\dfrac{(d+8)(d-2)}{4(d+6)}$ is the answer in reduced form.

d) $\dfrac{g}{5} - \dfrac{5}{g-24}$ Step 1: Make a new denominator by multiplying denominators together.

$\dfrac{g}{5(g-24)} - \dfrac{5}{5(g-24)}$ Step 2: Multiply the numerators by the same factor as the denominator.

$\dfrac{g(g-24)}{5(g-24)} - \dfrac{5 \bullet 5}{5(g-24)}$ Step 3: Use the distribution if possible.

$\dfrac{g^2 - 24g}{5(g-24)} - \dfrac{25}{5(g-24)}$ Step 4: combine like terms.

$\dfrac{g^2 - 24g - 25}{5(g-24)}$ Step 5: Factor out numerator if possible.

$\dfrac{(g-24)(g+1)}{5(g-24)}$ Step 6: Cancel any factors if possible.

$(g-24)$ cancels; therefore, $\dfrac{(g+1)}{5}$ is the answer in reduced form. ∎

Now let us take a look at fractions where the denominators are binomials.

EXAMPLE 6:

a) $\dfrac{9}{h+1} + \dfrac{1}{h+3}$ Step 1: Make a new denominator by multiplying denominators together.

$\dfrac{9}{(h+1)(h+3)} + \dfrac{1}{(h+1)(h+3)}$ Step 2: Multiply the numerators by the same factor as the denominator.

$\dfrac{9(h+3)}{(h+1)(h+3)} + \dfrac{1(h+1)}{(h+1)(h+3)}$ Step 3: Use the distribution if possible.

$$\frac{9h+27}{(h+1)(h+3)}+\frac{h+1}{(h+1)(h+3)}$$ Step 4: combine like terms.

$$\frac{10h+28}{(h+1)(h+3)}$$ Step 5: Factor out numerator if possible.

$$\frac{2(5h+14)}{(h+1)(h+3)}$$ Step 6: Cancel any factors if possible.

Nothing cancels; therefore, $\dfrac{2(5h+14)}{(h+1)(h+3)}$ is the answer in reduced form.

b) $\dfrac{4}{j-2}+\dfrac{1}{j+1}$ Step1: Make a new denominator by multiplying denominators together.

$$\frac{4}{(j-2)(j+1)}+\frac{1}{(j-2)(j+1)}$$ Step 2: Multiply the numerators by the same factor as the denominator.

$$\frac{4(j+1)}{(j-2)(j+1)}+\frac{1(j-2)}{(j-2)(j+1)}$$ Step 3: Use the distribution if possible.

$$\frac{4j+4}{(j-2)(j+1)}+\frac{j-2}{(j-2)(j+1)}$$ Step 4: combine like terms.

$$\frac{5j-2}{(j-2)(j+1)}$$ Step 5: Factor out numerator if possible.

$$\frac{5j-2}{(j-2)(j+1)}$$ Step 6: Cancel any factors if possible.

Nothing cancels; therefore, $\dfrac{5j-2}{(j-2)(j+1)}$ is the answer in reduced form.

c) $\dfrac{2}{k-3}-\dfrac{3}{k+2}$ Step1: Make a new denominator by multiplying denominators together.

$$\frac{2}{(k-3)(k+2)}-\frac{3}{(k-3)(k+2)}$$ Step 2: Multiply the numerators by the same factor as the denominator.

$$\frac{2(k+2)}{(k-3)(k+2)} - \frac{3(k-3)}{(k-3)(k+2)}$$ Step 3: Use the distribution if possible.

$$\frac{2k+4}{(k-3)(k+2)} - \frac{3k-9}{(k-3)(k+2)}$$ Step 4: combine like terms.

$$\frac{-k+13}{(k-3)(k+2)}$$ Step 5: Factor out numerator if possible.

$$\frac{-k+13}{(k-3)(k+2)}$$ Step 6: Cancel any factors if possible.

Nothing cancels; therefore, $\dfrac{-k+13}{(k-3)(k+2)}$ is the answer in reduced form.

d) $\dfrac{9}{m-1} - \dfrac{4}{m+2}$ Step1: Make a new denominator by multiplying denominators together.

$$\frac{9}{(m-1)(m+2)} - \frac{4}{(m-1)(m-2)}$$ Step 2: Multiply the numerators by the same factor as the denominator.

$$\frac{9(m+2)}{(m-1)(m+2)} - \frac{4(m-1)}{(m-1)(m+2)}$$ Step 3: Use the distribution if possible.

$$\frac{9m+18}{(m-1)(m+2)} - \frac{4m-4}{(m-1)(m+2)}$$ Step 4: combine like terms.

$$\frac{5m+22}{(m-1)(m+2)}$$ Step 5: Factor out numerator if possible.

$$\frac{5m+22}{(m-1)(m+2)}$$ Step 6: Cancel any factors if possible.

Nothing cancels; therefore, $\dfrac{5m+22}{(m-1)(m+2)}$ is the answer in reduced form. ∎

Recall that the minus sign is distributed through the numerator when combining like terms. The numerator was combined as 9m + 18 − (4m − 4) in example 6d.

I will end this section with an example that has trinomials involved.

EXAMPLE 7:

a) $\dfrac{5}{n^2 - 3n + 2} + \dfrac{4}{n-2}$ Step1: Factor denominators.

$\dfrac{5}{(n-2)(n-1)} + \dfrac{4}{n-2}$ Step 2: Make a new denominator by multiplying the binomial in the second fraction by the non-matching binomial from the first fraction.

$\dfrac{5}{(n-2)(n-1)} + \dfrac{4}{(n-2)(n-1)}$ Step 3: Multiply the numerators by the same factor as the denominator.

$\dfrac{5}{(n-2)(n-1)} + \dfrac{4(n-1)}{(n-2)(n-1)}$ Step 4: Use the distribution if possible.

$\dfrac{5}{(n-2)(n-1)} + \dfrac{4n-4}{(n-2)(n-1)}$ Step 5: combine like terms.

$\dfrac{4n+1}{(n-2)(n-1)}$ Step 6: Factor out numerator if possible.

$\dfrac{4n+1}{(n-2)(n-1)}$ Step 7: Cancel any factors if possible.

Nothing cancels; therefore, $\dfrac{4n+1}{(n-2)(n-1)}$ is the answer in reduced form.

b) $\dfrac{1}{p^2 - 5p + 6} + \dfrac{1}{p-2}$ Step1: Factor denominators.

$\dfrac{1}{(p-2)(p-3)} + \dfrac{1}{(p-2)}$ Step 2: Make a new denominator by multiplying the binomial in the second fraction by the non-matching binomial from the first fraction.

$\dfrac{1}{(p-2)(p-3)} + \dfrac{1}{(p-2)(p-3)}$ Step 3: Multiply the numerators by the same factor as the denominator.

$\dfrac{1}{(p-2)(p-3)} + \dfrac{1(p-3)}{(p-2)(p-3)}$ Step 4: Use the distribution if possible.

234

$$\frac{1}{(p-2)(p-3)}+\frac{p-3}{(p-2)(p-3)}$$ Step 5: combine like terms.

$$\frac{p-2}{(p-2)(p-3)}$$ Step 6: Factor out numerator if possible.

$$\frac{p-2}{(p-2)(p-3)}$$ Step 7: Cancel any factors if possible.

(p – 2) cancels; therefore, $\dfrac{1}{(p-3)}$ is the answer in reduced form.

c) $\dfrac{4}{q^2-16}-\dfrac{1}{q^2-q-12}$ Step1: Factor denominators.

$$\frac{4}{(q-4)(q+4)}-\frac{1}{(q-4)(q+3)}$$ Step 2: Make new denominators by multiplying the binomial in the second fraction by the non-matching binomial from the first fraction and the binomial in the first fraction by the non-matching binomial from the second fraction.

$$\frac{4}{(q-4)(q+4)(q+3)}-\frac{1}{(q-4)(q+3)(q+4)}$$ Step 3: Multiply the numerators by the same factor as the denominator.

$$\frac{4(q+3)}{(q-4)(q+4)(q+3)}-\frac{1(q+4)}{(q-4)(q+3)(q+4)}$$ Step 4: Use the distribution if possible.

$$\frac{4q+12}{(q-4)(q+4)(q+3)}-\frac{q+4}{(q-4)(q+3)(q+4)}$$ Step 5: combine like terms.

$$\frac{3q+8}{(q-4)(q+4)(q+3)}$$ Step 6: Factor out numerator if possible.

$$\frac{3q+8}{(q-4)(q+4)(q+3)}$$ Step 7: Cancel any factors if possible.

Nothing cancels; therefore, $\dfrac{3q+8}{(q-4)(q+4)(q+3)}$ is the answer in reduced form.

d) $\dfrac{5r}{r^2+r-6} - \dfrac{1}{r^2-2r-15}$ Step1: Factor denominators.

$\dfrac{5r}{(r-2)(r+3)} - \dfrac{1}{(r-5)(r+3)}$ Step 2: Make new denominators by multiplying the binomial in the second fraction by the non-matching binomial from the first fraction and the binomial in the first fraction by the non-matching binomial from the second fraction.

$\dfrac{5r}{(r-2)(r+3)(r-5)} - \dfrac{1}{(r-5)(r+3)(r-2)}$ Step 3: Multiply the numerators by the same factor as the denominator.

$\dfrac{5r(r-5)}{(r-2)(r+3)(r-5)} - \dfrac{1(r-2)}{(r-5)(r+3)(r-2)}$ Step 4: Use the distribution if possible.

$\dfrac{5r^2-25r}{(r-2)(r+3)(r-5)} - \dfrac{r-2}{(r-5)(r+3)(r-2)}$ Step 5: combine like terms.

$\dfrac{5r^2-26r+2}{(r-2)(r+3)(r-5)}$ Step 6: Factor out numerator if possible.

$\dfrac{5r^2-26r+2}{(r-2)(r+3)(r-5)}$ Step 7: Cancel any factors if possible.

Nothing cancels; therefore, $\dfrac{5r^2-26r+2}{(r-2)(r+3)(r-5)}$ is the answer in reduced form. ∎

It is not that uncommon to have seven steps in solving algebraic fractions. More than two binomials are needed in the denominators when trinomials are involved as was shown in Example 7.

PROBLEMS:

Evaluate:

1) $\dfrac{7}{8} + \dfrac{5}{6}$

2) $\dfrac{5}{8} + \dfrac{3}{5}$

3) $\dfrac{7}{9} - \dfrac{1}{6}$

4) $\dfrac{11}{15} - \dfrac{3}{10}$

5) $\dfrac{3s}{4} + \dfrac{s}{9}$

6) $\dfrac{t}{4} + \dfrac{3t}{5}$

7) $\dfrac{5u}{6} - \dfrac{2u}{3}$

8) $\dfrac{7v}{3} + \dfrac{v}{7}$

9) $\dfrac{1}{w} + \dfrac{1}{x}$

10) $\dfrac{5}{y} + \dfrac{8}{z}$

11) $\dfrac{2}{3a} - \dfrac{2}{7b}$

12) $\dfrac{1}{5c} - \dfrac{1}{2d}$

13) $\dfrac{5}{h} + \dfrac{1}{5}$

14) $\dfrac{5}{f} + \dfrac{2}{3}$

15) $\dfrac{3}{g} - \dfrac{4}{5}$

16) $\dfrac{1}{3} - \dfrac{3}{h}$

17) $\dfrac{j+1}{5} + \dfrac{j+7}{2}$

18) $\dfrac{k+2}{2} + \dfrac{k+3}{4}$

19) $\dfrac{m-1}{5} - \dfrac{m-4}{9}$

20) $\dfrac{n-2}{3} - \dfrac{n-1}{2}$

21) $\dfrac{p+3}{2} + \dfrac{4p-1}{6}$

22) $\dfrac{q+1}{7} - \dfrac{4q-1}{10}$

23) $\dfrac{5}{r} + \dfrac{3}{r^2}$

24) $\dfrac{1}{s} + \dfrac{2}{s^3}$

25) $\dfrac{4}{t^2} - \dfrac{3}{t}$

26) $\dfrac{1}{u} - \dfrac{2}{u^3}$

27) $\dfrac{v+2}{v^4} + \dfrac{v-1}{v^7}$

28) $\dfrac{w+9}{w^6} + \dfrac{w-2}{w^3}$

29) $\dfrac{x+1}{x^2} - \dfrac{3x-2}{x^4}$

30) $\dfrac{y+1}{y^8} - \dfrac{5y-2}{y^5}$

31) $\dfrac{4}{z} + \dfrac{3}{z+1}$

32) $\dfrac{4}{a+2} + \dfrac{3}{a}$

33) $\dfrac{2}{b} - \dfrac{1}{b-2}$

34) $\dfrac{5}{c-1} - \dfrac{2}{c}$

35) $\dfrac{d}{d+2} + \dfrac{2}{5}$

36) $\dfrac{h}{h+3} + \dfrac{2}{3}$

37) $\dfrac{f}{f-4} - \dfrac{3}{4}$

38) $\dfrac{3}{4} - \dfrac{g}{g-1}$

39) $\dfrac{2}{h+1} + \dfrac{3}{h+3}$

40) $\dfrac{5}{j-1} + \dfrac{2}{j+2}$

41) $\dfrac{2}{k+3} + \dfrac{1}{k+1}$

42) $\dfrac{2}{m+3} + \dfrac{1}{m+2}$

43) $\dfrac{4}{n-2} - \dfrac{1}{n+1}$

44) $\dfrac{5}{p+4} - \dfrac{3}{p-1}$

45) $\dfrac{2}{q+3} - \dfrac{1}{q+1}$

46) $\dfrac{2}{r+3} - \dfrac{1}{r+2}$

47) $\dfrac{2}{s^2-3s+2} + \dfrac{3}{s-2}$

48) $\dfrac{1}{t^2-5t+6} + \dfrac{1}{t-2}$

49) $\dfrac{3}{u^2-u-12} + \dfrac{1}{u-4}$

50) $\dfrac{2}{v^2+v-12} + \dfrac{4}{v+4}$

51) $\dfrac{5}{w^2-16} - \dfrac{3}{w^2-w-12}$

52) $\dfrac{3}{x^2+4x+3} - \dfrac{1}{x^2-9}$

53) $\dfrac{2}{y^2+y-6} - \dfrac{7}{y^2-2y-15}$

54) $\dfrac{3}{z^2-z-12} - \dfrac{2}{z^2-2z-8}$

SECTION 9.6
COMPLEX FRACTIONS

 Complex fractions are fractions within fractions. A complex fraction in its simplest form is a fraction on top of a fraction. A fraction is another form of division where the numerator is divided by the denominator. Example 1 will show a numeric complex fraction and an algebraic complex fraction.

EXAMPLE 1:

a) $\dfrac{\frac{2}{3}}{\frac{1}{2}}$ Step1: Change to a division problem.

$\dfrac{2}{3} \div \dfrac{1}{2}$ Step 2: Take the reciprocal of the right fraction so that the division problem turns into a multiplication problem.

$\dfrac{2}{3} \bullet \dfrac{2}{1}$ Step 3: Multiply the fractions.

$\dfrac{4}{3}$ Step 4: Reduce the answer if possible.

The fraction is not reducible; therefore, $\dfrac{4}{3}$ is the answer in reduced form.

b) $\dfrac{\frac{a}{2}}{\frac{3}{b}}$ Step1: Change to a division problem.

$\dfrac{a}{2} \div \dfrac{3}{b}$ Step 2: Take the reciprocal of the right fraction so that the division problem turns into a multiplication problem.

$\dfrac{a}{2} \bullet \dfrac{b}{3}$ Step 3: Multiply the fractions.

$\dfrac{ab}{6}$ Step 4: Reduce the answer if possible.

The fraction is not reducible; therefore, $\dfrac{ab}{6}$ is the answer in reduced form.■

Notice that the same rules from basic mathematics were applied to the algebraic fractions. Once again the rules have not changed. Now let us look at an example where powers and factoring are involved.

EXAMPLE 2:

a) $\dfrac{\dfrac{3}{c^3}}{\dfrac{2}{c^2}}$ Step1: Change to a division problem.

$\dfrac{3}{c^3} \div \dfrac{2}{c^2}$ Step 2: Take the reciprocal of the right fraction so that the division problem turns into a multiplication problem.

$\dfrac{3}{c^3} \bullet \dfrac{c^2}{2}$ Step 3: Multiply the fractions.

$\dfrac{3c^2}{2c^3}$ Step 4: Reduce the answer if possible.

The fraction is reducible; therefore, $\dfrac{3}{2c}$ is the answer in reduced form.

b) $\dfrac{\dfrac{d+2}{d^2+3d+2}}{\dfrac{d^2-4}{d^2+12d+20}}$ Step1: Change to a division problem.

$\dfrac{d+2}{d^2+3d+2} \div \dfrac{d^2-4}{d^2+12d+20}$ Step 2: Take the reciprocal of the right fraction so that the division problem turns into a multiplication problem.

$\dfrac{d+2}{d^2+3d+2} \bullet \dfrac{d^2+12d+20}{d^2-4}$ Step 3: Factor the fractions.

$\dfrac{d+2}{(d+1)(d+2)} \bullet \dfrac{(d+2)(d+10)}{(d+2)(d-2)}$ Step 4: Cancel where possible and multiply.

d + 2 cancels; therefore, $\dfrac{d+10}{(d+1)(d-2)}$ is the answer in reduced form.

Now let us look at an example where sums and differences are part of a complex fraction.

EXAMPLE 3:

a) $\dfrac{2+\dfrac{f}{g}}{3-\dfrac{f}{g}}$ Step1: Remove fractions with sum and differences by multiplying through by g.

$\dfrac{2g+f}{3g-f}$ Step 2: Factor and reduce if possible.

There is nothing factorable; therefore, $\dfrac{2g+f}{3g-f}$ is the answer in reduced form.

b) $\dfrac{4-\dfrac{j^2}{k^2}}{1+\dfrac{j}{k}}$ Step1: Remove fractions with sum and differences by multiplying through by k^2.

$\dfrac{4k^2-j^2}{k^2+jk}$ Step 2: Factor and reduce if possible.

$\dfrac{(2k+j)(2k-j)}{k(k+j)}$

Nothing cancels; therefore, $\dfrac{(2k+j)(2k-j)}{k(k+j)}$ is the answer in reduced form.■

Notice that the complex fraction became simpler to work with when the fractions were divided by something that will cancel the fraction in the sum and difference. The variable with the highest exponent will always eliminate the denominators as was the case in example 3b.

PROBLEMS:

Evaluate:

1) $\dfrac{\dfrac{2}{3}}{\dfrac{5}{12}}$

2) $\dfrac{\dfrac{3}{5}}{\dfrac{2}{7}}$

3) $\dfrac{\dfrac{4}{l}}{\dfrac{2}{m}}$

4) $\dfrac{\dfrac{8}{p}}{\dfrac{7}{q}}$

5) $\dfrac{\dfrac{r}{8}}{\dfrac{r^2}{4}}$

6) $\dfrac{\dfrac{3}{s}}{\dfrac{2}{s^2}}$

7) $\dfrac{\dfrac{t+1}{t^2+2t+1}}{\dfrac{t^2-4}{t^2+t-2}}$

8) $\dfrac{\dfrac{u-7}{u^2-2u-3}}{\dfrac{u^2-9}{u^2-9u+14}}$

9) $\dfrac{\dfrac{v^2-1}{v^2+2v+1}}{\dfrac{v^2-2v-8}{v-3}}$

10) $\dfrac{\dfrac{w+5}{w^2-16}}{\dfrac{w+7}{w^2-t-12}}$

11) $\dfrac{2-\dfrac{1}{x}}{2+\dfrac{1}{x}}$

12) $\dfrac{3+\dfrac{1}{y}}{3-\dfrac{1}{y}}$

13) $\dfrac{z^2-1}{1-\dfrac{1}{z}}$

14) $\dfrac{\dfrac{a}{b}+2}{\dfrac{a^2}{b^2}-4}$

SECTION 9.7
SOLVING ALGEBRAIC EQUATIONS WITH FRACTIONS

Solving algebraic equations with fractions is almost the same as solving regular and quadratic algebraic equations. The only differences are that the fractions must be eliminated and solutions may not exist. A solution does not exist when an answer sends the denominator to zero. Refer back to section 9.1 example 4 on page 208.

EXAMPLE 1:

Solve for the unknown.

a) $\dfrac{a}{9} = 36$ Step 1: Multiply entire equation to eliminate fractions.

$9 \bullet \dfrac{a}{9} = 36 \bullet 9$ Step 2: Simplify equation.

a = 324 Step 3: Check answer in denominator to see if denominator goes to zero.

The unknown is not in the denominator; therefore, a = 324.

It is left up to the reader to check the answer.

b) $\dfrac{7}{b} = 45$ Step 1: Multiply entire equation to eliminate fractions.

$b \bullet \dfrac{7}{b} = 45b$ Step 2: Simplify equation.

7 = 45b Step 3: Solve for unknown.

$\dfrac{7}{45} = \dfrac{45b}{45}$ Step 5: Simplify equation.

$\dfrac{1}{5} = b$ Step 6: Check answer in denominator to see if denominator goes to zero.

The unknown in the denominator does not go to zero; therefore, $\dfrac{1}{5} = b$.■

There will always be a solution when no variables are in the denominators. Zero is the only number that will cause no solution when there is only an unknown in the denominator. In example 1b, b=0 would cause the equation to have no solution, but since there was an answer, a solution did exist.

Now let us look at algebraic fractions where there are more than one term in the denominator.

EXAMPLE 2:

Solve for the unknown.

a) $\dfrac{2}{c+3} = 3$ Step 1: Multiply entire equation to eliminate fractions.

$(c+3)\left(\dfrac{2}{c+3}\right) = 3(c+3)$ Step 2: Simplify equation.

$2 = 3c + 9$ Step 3: Isolate the variable.
$-9 \qquad -9$

$-7 = 3c$ Step 4: Solve for the unknown

$\dfrac{-7}{3} = c$ Step 5: Check answer in denominator to see if denominator goes to zero.

The unknown in the denominator does not go to zero; therefore, $\dfrac{-7}{3} = c$.

b) $\dfrac{9}{d^2 - 4} = 4$ Step 1: Multiply entire equation to eliminate fractions.

$(d^2 - 4)\left(\dfrac{9}{d^2 - 4}\right) = 4(d^2 - 4)$ Step 2: Simplify equation.

$9 = 4d^2 - 16$ Step 3: Set quadratic equation to zero.

$0 = 4d^2 - 25$ Step 4: Solve quadratic equation by factoring.

$0 = (2d + 5)(2d - 5)$ Step 6: Solve for the unknown.

$$0 = 2d + 5 \qquad 0 = 2d - 5$$
$$-5 = 2d \qquad\quad 5 = 2d$$
$$\frac{-5}{2} = d \qquad\quad \frac{5}{2} = d$$

Step 7: Check answer in denominator to see if denominator goes to zero.

The unknown in the denominator does not go to zero; therefore, $\frac{-5}{2} = d$ or $\frac{5}{2} = d$. ∎

An alternative method would be to set the denominators to zero and see what answers would make the denominator go to zero and then solve equations to see if the solutions match the ones that make the denominator go to zero. A solution of 2 or -2 would cause the denominator to go to zero in example 2b. Clearly, the solutions to example 2b do not match the numbers to make the denominator go to zero.

Finally, let us look at some equations that have sums and differences in them.

EXAMPLE 3:

a) $\quad \dfrac{f}{2} + \dfrac{1}{3} = \dfrac{2f+3}{6}$ Step 1: Multiply entire equation to eliminate fractions.

$$6\left(\frac{f}{2}\right) + 6 \bullet \frac{1}{3} = 6\left(\frac{2f+3}{6}\right) \text{ Step 2: Simplify equation.}$$

$3f + 2 = 2f + 3$ Step 3: Solve for the unknown.

$$\begin{array}{r} 3f + 2 = 2f + 3 \\ -2f - 2 \quad -2f - 2 \end{array}$$

$f = 1$ Step 5: Check answer in denominator to see if denominator goes to zero.

The unknown is not in the denominator; therefore, $f = 1$.

b) $\quad \dfrac{g}{g-3} - 2 = \dfrac{1}{g-3}$ Step 1: Multiply entire equation to eliminate fractions.

$$(g-3)\left(\frac{g}{g-3}\right) - 2(g-3) = (g-3)\left(\frac{1}{g-3}\right) \text{ Step 2: Simplify equation.}$$

$g - 2g + 6 = 1$ Step 3: Solve for the unknown.

$-g + 6 = 1$
  $- 6 \ - 6$

$-g = -5$

$g = 5$ Step 4: Check answer in denominator to see if denominator goes to zero.

The unknown in the denominator does not go to zero; therefore, $g = 5$.

c) $\dfrac{h}{h-4} = \dfrac{15}{h-3} - \dfrac{2h}{h^2 - 7h + 12}$ Step 1: Factor trinomial.

$\dfrac{h}{h-4} = \dfrac{15}{h-3} - \dfrac{2h}{(h-4)(h-3)}$ Step 2: Eliminate fractions.

$(h-4)(h-3)\left(\dfrac{h}{h-4}\right) = (h-4)(h-3)\left(\dfrac{15}{h-3}\right) - (h-4)(h-3)\dfrac{2h}{(h-4)(h-3)}$

Step 3: Simplify equation.

$(h-3)h = (h-4)15 - 2h$
$h^2 - 3h = 15h - 60 - 2h$ Step 4: Solve the quadratic equation.

$h^2 - 16h + 60 = 0$

$(h-6)(h-10) = 0$

$h = 6$ or $h = 10$ Step 5: Check answer in denominator to see if denominator goes to zero.

The unknown in the denominator does not go to zero; therefore, $h = 6$ or $h = 10$.∎

Notice that the fractions were multiplied by a factor that eliminated all the fractions. One could eliminate one fraction at a time, but this method would cause more work.

SUMMARY: Eliminate the fractions.
Solve for the unknown.
Check solution in denominator.

PROBLEMS:

Solve for the unknown

1) $\dfrac{j}{2} = 4$

2) $\dfrac{k}{12} = 40$

3) $\dfrac{2}{l} = 4$

4) $\dfrac{9}{m} = 2$

5) $\dfrac{3}{n+9} = 4$

6) $\dfrac{1}{p-3} = 44$

7) $\dfrac{9}{q^2 - 4} = 4$

8) $\dfrac{25}{r^2 - 16} = 9$

9) $\dfrac{s}{5} - \dfrac{1}{3} = \dfrac{s-7}{3}$

10) $\dfrac{t}{6} + \dfrac{3}{4} = \dfrac{t-1}{4}$

11) $\dfrac{u}{u-3} - 2 = \dfrac{3}{u-3}$

12) $\dfrac{12}{v+3} = \dfrac{v}{v+3} + 2$

13) $\dfrac{5}{w-4} = \dfrac{1}{w+2} - \dfrac{2}{w^2 - 2w - 8}$

14) $\dfrac{3}{x-1} - \dfrac{1}{x+9} = \dfrac{18}{x^2 + 8x - 9}$

15) $\dfrac{3}{y+3} + \dfrac{25}{y^2 + y - 6} = \dfrac{5}{y-2}$

16) $\dfrac{7}{z-5} - \dfrac{3}{z+5} = \dfrac{40}{z^2 - 25}$

17) $\dfrac{1}{a+4} + \dfrac{1}{a-4} = \dfrac{12}{a^2 - 16}$

18) $\dfrac{11}{b+2} = \dfrac{5}{b^2 - b - 6} + \dfrac{1}{b-3}$

19) $\dfrac{2}{c+2} = \dfrac{3}{c+6} + \dfrac{9}{c^2 + 8c + 12}$

20) $\dfrac{5}{d+6} + \dfrac{2}{d^2 + 7d + 6} = \dfrac{3}{d+1}$

21) $\dfrac{3}{f-3} - \dfrac{18}{f^2 - 9} = \dfrac{5}{f+3}$

22) $\dfrac{2g}{g-3} + \dfrac{2}{g-5} = \dfrac{3g}{g^2 - 8g + 15}$

23) $\dfrac{h}{h-4} - \dfrac{5h}{h^2 - h - 12} = \dfrac{3}{h+3}$

24) $\dfrac{2j}{j+2} = \dfrac{5}{j^2 - j - 6} - \dfrac{1}{j-3}$

SECTION 9.8
RATIO & PROPORTION

Proportions show that two fractions are equal. Sometimes a missing numerator or denominator needed to be found. Recall from basic mathematics where two fractions were separated by an equal sign. One fraction had a missing numerator or denominator. Cross multiplication was used along with division to solve for the missing numerator or denominator.

EXAMPLE 1:

a) $\dfrac{2}{5} = \dfrac{a}{20}$ Step 1: Cross multiply.

$5a = 2 \bullet 20$ Step 2: Solve for unknown.

$5a = 40$
$a = 8$

$\dfrac{2}{5} = \dfrac{8}{20}$.

b) $\dfrac{4}{11} = \dfrac{12}{b}$ Step 1: Cross Multiply.

$11 \bullet 12 = 4b$ Step 2: Solve for the unknown.

$132 = 4b$
$33 = b$

$\dfrac{4}{11} = \dfrac{12}{33}$

It is left to the reader to check the solutions.∎

Cross multiplication can also help solve proportions where the are unknowns in more than one place in the two fractions.

EXAMPLE 2:

a) $\dfrac{c-1}{9} = \dfrac{c+2}{12}$ Step 1: Cross multiply.

$9(c + 2) = 12(c - 1)$ Step 2: Solve for unknown.

$$9c + 18 = 12c - 12$$
$$9c - 12c = -12 - 18$$
$$-3c = -30$$
$$c = 10$$

There is no need to check to see if the denominator goes to zero since the unknown is in the numerator.

b)　　$\dfrac{5}{d-3} = \dfrac{15}{21}$ Step 1: Cross Multiply.

$15(d - 3) = 5 \bullet 21$ Step 2: Solve for the unknown.

$$15d - 45 = 105$$
$$15d = 150$$
$$d = 10$$

The denominator does not go to zero; therefore, d − 10 is the solution. ∎

A ratio shows that two things with different units are the same. 1 gallon of milk equals 4 quarts of milk would be a ratio. The same cross multiplication is used to solve ratios.

EXAMPLE 3:

a)　　How many ounces are in 5 pounds when 1 pound equals 16 ounces? Step 1: Set up

a proportion.

$\dfrac{1lb}{16oz} = \dfrac{5lb}{?\,oz}$ Step 2: Cross multiply.

$16oz \bullet 5lb = 1lb\ ?oz$. Step 3: Solve for unknown.

$80oz\ lb = 1lb\ ?oz$
$80 = ?$

5 pounds equals 80 ounces.

b) How many feet are in 42 inches where 1 foot equals 12 inches? Step 1: Set up a proportion.

$$\frac{1ft}{12in} = \frac{?\,ft}{42in}$$ Step 2: Cross multiply.

$12in\ ?ft = 1ft\ 42in.$ Step 3: Solve for unknown.

$12in\ ?ft = 42in\,ft$
$3.5 = ?$

3.5 feet equals 42 ounces.■

I used a question mark instead of a variable because it could be hard to separate the unknown from the unit of measure. The units of measures cancel as if they were variables and that is why there is no unit of measure in the solution. The solution is substituted for the question mark and the unit of measure is then established for the solution.

SUMMARY: Cross Multiply.
Solve for the unknown.
Check solution in denominator if possible.

PROBLEMS:

Solve for the unknown

1) $\dfrac{f}{7} = \dfrac{8}{14}$　　　　　　2) $\dfrac{3}{g} = \dfrac{9}{15}$

3) $\dfrac{5}{8} = \dfrac{20}{h}$　　　　　　4) $\dfrac{j}{10} = \dfrac{9}{30}$

5) $\dfrac{k+1}{5} = \dfrac{20}{25}$　　　　6) $\dfrac{2}{5} = \dfrac{l-2}{20}$

7) $\dfrac{m}{6} = \dfrac{m+5}{16}$　　　　8) $\dfrac{n-2}{n+2} = \dfrac{12}{20}$

9) $\dfrac{2}{p-1} = \dfrac{6}{p+9}$　　　10) $\dfrac{3}{q-3} = \dfrac{4}{q-5}$

Use the following ratios to solve the following problems.

1pound = 16 ounces 1 gallon = 4 quarts 1 foot = 12 inches 1 mile = 5280 feet
60 minutes = 1 hour 60 seconds = 1 minute 3 feet = 1 yard

11) How many ounces are in 24 pounds?

12) How many pounds are in 12 ounces?

13) How many gallons are in 24 quarts?

14) How many quarts are in 44 gallons?

15) How many feet are in 54 inches?

16) How many inches are in 38 feet?

17) How many miles are in 15840 feet?

18) How many feet are in 2 miles?

19) How many minutes are in 4 hours?

20) How many hours are in 12 minutes?

21) How many minutes are in 48 seconds?

22) How many seconds are in 25 minutes?

23) How many feet are in 26 yards?

24) How many yards are in 33 feet?

Two conversions are needed to solve the following problems.

25) How many yards are in 1 mile?

26) How many seconds are in 12 hours?

27) How many hours are in 252 seconds?

28) How many hours are in 540 seconds?

SECTION 9.9
SOLVING WORD PROBLEMS WITH FRACTIONS

Let us now applied what was learned to solving word problems.

EXAMPLE 1:

If one-third of a number is added to three-fourths of that same number, the sum is 52. Find the number.

Step 1: Translate the problem into an algebraic equation.

$\frac{1}{3}x + \frac{3}{4}x = 52$ Step 2: Multiply through to eliminate fractions.

$12 \bullet \frac{1}{3}x + 12 \bullet \frac{3}{4}x = 52 \bullet 12$ Step 3: Simplify equation.

$4x + 9x = 624$. Step 4: Solve for unknown.

$13x = 624$
$x = 48$.

It is left to the reader to check the solution.■

EXAMPLE 2:

One number is twice another number. If their sum of their reciprocals is ½, what are the two numbers?

Step 1: Translate the problem into an algebraic equation.

$\frac{1}{x} + \frac{1}{2x} = \frac{1}{2}$ Step 2: Multiply through to eliminate fractions.

$2x \bullet \frac{1}{x} + 2x \bullet \frac{1}{2x} = \frac{1}{2} \bullet 2x$ Step 3: Simplify equation.

$2 + 1 = x$.   Step 4: Solve for unknown.

$3 = x$ and $2x = 6$■

EXAMPLE 3:

  Velma took one hour longer to drive 200 miles than she did on a trip of 160 miles. If her speed was the same both times, how long did each trip take?

STEP 1: Create a chart for distance(D), rate(R), and time (T)

	Distance	Time	Rate
200 mile trip	200	$T+1$	$\dfrac{200}{T+1}$
160 mile trip	160	T	$\dfrac{160}{T}$

STEP 2: Set up algebraic equation.

$\dfrac{200}{T+1} = \dfrac{160}{T}$ STEP 3: Solve for unknown.

$200T = 160T + 160$
$40T = 160$
$T = 4$

The 200 mile trip took 5 hours and the 160 mile trip took four hours. ■

EXAMPLE 4:

  A train makes a trip of 225 miles in the same time a bus can travel 187.5 miles. If the speed of the train is 10 miles per hour faster than the speed of the bus, find the speed of each.

STEP 1: Create a chart for distance(D), rate(R), and time (T)

	Distance	Rate	Time
Train	225	$R+10$	$\dfrac{225}{R+10}$
Bus	187.5	R	$\dfrac{187.5}{R}$

STEP 2: Set up algebraic equation.

$\dfrac{225}{R+10} = \dfrac{187.5}{R}$ STEP 3: Solve for unknown.

$225R = 187.5R + 1875$
$37.5R = 1875$
$R = 50$

The bus is traveling at a speed of 50 miles per hour and the train is traveling at a speed of 60 miles per hour.■

EXAMPLE 5:

A car uses 2 gallons of gas to travel 110 miles. At that mileage rate, how many gallons will be used on a trip of 385 miles?

STEP 1: Set up a proportion.

$\dfrac{2 gallons}{110 miles} = \dfrac{x gallons}{385 miles}$ STEP 2: Solve for unknown.

770 gallons miles = 110x gallons miles.
7 = x.

The car will use 7 gallons of gas to travel 385 miles.■

Remember to set the proportion with the same units in the denominator as well as the numerator.

PROBLEMS:

1)   The sum of two-fifths of a number and one-half a number is 18. Find the number.

2)   If one-third of a number is subtracted from three-fourths of that number, the difference is 15. What is the number?

3)   One number is three times another number. If the sum of their reciprocals is $\dfrac{2}{9}$, find the two numbers.

4)   One number is twice another number. If the sum of their reciprocals is ¼, find the two numbers.

5)   Carla took one hour longer to bicycle 60 miles than she did on a trip of 45 miles. If her speed was the same each time, find the time for each trip.

6)   A passenger train can travel 325 miles in the same time a freight train takes to travel 200 miles. If the speed of the passenger train is 25 miles per hour faster than the freight, find the speed of each.

7)   A car makes a trip of 280 miles in the same time a truck travels 245 miles, Of the speed of the truck is 5 miles per hour faster than that of the car, find the speed of each.

8)   A light plane took one hour longer to travel 450 miles on the first portion of a trip than it took to fly 300 miles on the second. If the speed was the same for each portion, what was the flying time for each part of the trip?

9)   A car uses 8 liters of gasoline in traveling 100 kilometers. At that rate, how many liters of gas will be used on a trip of 250 kilometers?

10)   Jill completed 24 pages of her reading assignment in 60 minutes. If the assignment is 100 pages long, how long will the entire assignment take if Jill reads at the same rate?

11)   A car uses 5 gallons of gasoline on a trip of 160 miles. At the same mileage rate, how much gasoline will a 384 mile trip require?

12)   A car uses 12 liters of gasoline in traveling 150 kilometers. At that rate, how many liters of gasoline will be used on a trip of 400 kilometers?

13)   Shirley earns $6500 commission in 20 weeks in her new sales position. At that rate, how much will she earn in 1 year which is 52 weeks?

14)   Ken earned $165 interest for 1 year on an investment of $1500. At the same rate, what amount of interest would be earned by an investment of $2500?

15)   If two-fifths of a number is added to one-half of that number, the sum is 27. Find the number.

16)   One number is 3 times another. If the sum of their reciprocals is ⅓, what are the two numbers?

17)   Ralph made a trip of 240 miles. Returning by a different route, he found that the distance was only 200 miles, but traffic slowed his speed down by 8 miles per hour. If the trip took the same time in both directions, what was Ralph's rate each way?

18)   On the first day of a vacation trip, Joan drove 225 miles. On the second day it took her 1 hour longer to drive 270 miles. If her average speed was the same both days, how long did she drive each day?

CHAPTER 10
MORE ON RADICALS AND EXPONENTS

SECTION 10.1
INTORDUCTION TO COMPLEX NUMBERS

Complex numbers are the results when a square root is taken from a negative number. By definition $\sqrt{-1} = i$. These results are not real and are considered imaginary. The lower case i is used to represent the imaginary numbers.

Let us look at how exponents affect i before looking at examples. We already saw the first power of i. $i^2 = -1$, $i^3 = -i$, and $i^4 = 1$.

PROOF:

For $i^2 = -1$

$i = \sqrt{-1}$ Square both sides.
$i^2 = -1$

For $i^3 = -i$

$i = \sqrt{-1}$ Square both sides.
$i^2 = -1$ Multiply both sides by i.
$i\, i^2 = -1i$
$i^3 = -i$

For $i^4 = 1$.

$i = \sqrt{-1}$ Square both sides.
$i^2 = -1$ Multiply both sides by i^2.
$i^2 i^2 = -1\, i^2$
$i^4 = (-1)(-1)$
$i^4 = 1\blacksquare$

EXAMPLE 1:

a) $\sqrt{-25} = 5i$

b) $\sqrt{-2} = i\sqrt{2}\ \blacksquare$

Just take the square root of a number and put i after the number. i goes in front of the radical when the square root does not come out evenly. A solution is easier to read when i is in front of the radical.

i can be treated like a variable when adding, subtracting or dividing.

EXAMPLE 2:

a) $2i + 5i = 7i$

b) $6i - 7i = -i$

c) $21i \div 3i = 7\blacksquare$

There is an extra step when imaginary numbers are multiplied.

EXAMPLE 3:

a) $4i \bullet 11i = 44i^2 = -44$

b) $(2 + i)(2 - i) = 4 - 2i + 2i - i^2 = 4 + 1 = 5$

c) $(5 + i)(2 - i) = 10 - 5i + 2i - i^2 = 10 - 3i + 1 = 11 - 3i$

d) $2(5 + i) = 10 + 2i\blacksquare$

i^2 always gets multiplied when it is attached. The double negative rule was used for $-(-1)$.

PROBLEMS:

Simplify

1) $\sqrt{-4}$ 2) $\sqrt{-16}$

3) $\sqrt{-51}$ 4) $\sqrt{-22}$

5) $25i + 11i$ 6) $51i + 57i$

7) $25i - 11i$ 8) $12i - 58i$

9) $25i \div 5i$ 10) $12i \div 3i$

11) $2i \bullet 5i$ 12) $i \bullet 8i$

13) $(5 + i)(5 - i)$ 14) $(1 + i)(1 - i)$

15) $(2 + i)(1 - i)$ 16) $(5 + i)(8 - i)$

17) $8(9 + 4i)$ 18) $2(3 - i)$

SECTION 10.2
ANOTHER METHOD TO SOLVING QUADRATIC EQUATIONS

It was shown that some quadratic equations have no solutions back in section 8.6 page 193. A solution did not exist if the trinomial could not be factored. There is a formula that can be used to solve any type of quadratic equation. The formula is called the quadratic formula and looks like this: $x = \dfrac{-b \pm \sqrt{b^2 - 4ac}}{2a}$. The general form of a quadratic equation is $ax^2 + bx + c = 0$. The numbers that are in places a, b, and c are placed into the quadratic formula to generate solutions. There are two types of solutions. A quadratic equation may have either real or imaginary solutions. A real solution will occur if $b^2 - 4ac$ is equal to a positive number. The square root of a positive number is always a real number. An imaginary solution will occur if $b^2 - 4ac$ is equal to a negative number. The square root of a negative number is always an imaginary number.

Let us revisit some examples from section 8.6 to see if we get the same results using the quadratic formula.

EXAMPLE 1:

    a)   Solve for x: $h^2 + h = 0$

$a = 1$, $b = 1$, and $c = 0$. c automatically equals 0 when there is no c in the quadratic equation.

Step 1: Place a, b and c into quadratic formula. $h = \dfrac{-1 \pm \sqrt{1^2 - 4 \bullet 1 \bullet 0}}{2 \bullet 1}$

Step 2: Simplify using order of operations. Take power first inside radical.

$h = \dfrac{-1 \pm \sqrt{1 - 4 \bullet 1 \bullet 0}}{2 \bullet 1}$. Multiply inside radical.

$h = \dfrac{-1 \pm \sqrt{1 - 0}}{2 \bullet 1}$ Subtract inside radical.

$h = \dfrac{-1 \pm \sqrt{1}}{2 \bullet 1}$ Take square root.

$h = \dfrac{-1 \pm 1}{2 \bullet 1}$ Multiply in the denominator.

$h = \dfrac{-1 \pm 1}{2}$ Separate numerator.

$h = \dfrac{-1 + 1}{2}$ or $h = \dfrac{-1 - 1}{2}$ Add numerators and reduce fraction if possible.

$$h = \frac{0}{2} = 0 \ \text{ or } \ h = \frac{-2}{2} = -1$$

The solutions are 0 and -1 which are the same solutions from section 8.6.

b)   Solve for n: $n^2 = 16$
Set equation to zero.
$n^2 - 16 = 0$
$a = 1$, $b = 0$, and $c = -16$. b automatically equals 0 when there is no b in the quadratic equation.

Step 1: Place a, b and c into quadratic formula. $n = \dfrac{0 \pm \sqrt{0^2 - 4 \bullet 1(-16)}}{2 \bullet 1}$

Step 2: Simplify using order of operations. Take power first inside radical.
$n = \dfrac{0 \pm \sqrt{0 - 4 \bullet 1(-16)}}{2 \bullet 1}$. Multiply inside radical.

$n = \dfrac{0 \pm \sqrt{0 + 64}}{2 \bullet 1}$ Add inside radical.

$n = \dfrac{0 \pm \sqrt{64}}{2 \bullet 1}$ Take square root.

$n = \dfrac{0 \pm 8}{2 \bullet 1}$ Multiply in the denominator.

$n = \dfrac{0 \pm 8}{2}$ Separate numerator.

$n = \dfrac{0 + 8}{2}$ or $n = \dfrac{0 - 8}{2}$ Add numerators and reduce fraction if possible.

$n = \dfrac{8}{2} = 4$ or $n = \dfrac{-8}{2} = -4$
The solutions are 4 and -4. ■

Both quadratic equations have the same result as they did back in section 8.6. The quadratic formula generates solutions without factoring. Now let us look at one more example before revisiting two more examples from section 8.6.

EXAMPLE 2:

Solve for u: $4u^2 + 11u = -6$

Set equation to zero.
$4u^2 + 11u + 6 = 0$
$a = 4$, $b = 11$, and $c = 6$.

Step 1: Place a, b and c into quadratic formula. $u = \dfrac{-11 \pm \sqrt{11^2 - 4 \bullet 4 \bullet 6}}{2 \bullet 4}$

Step 2: Simplify using order of operations. Take power first inside radical.

$u = \dfrac{-11 \pm \sqrt{121 - 4 \bullet 4 \bullet 6}}{2 \bullet 4}$. Multiply inside radical.

$u = \dfrac{-11 \pm \sqrt{121 - 96}}{2 \bullet 4}$ Subtract inside radical.

$u = \dfrac{-11 \pm \sqrt{25}}{2 \bullet 4}$ Take square root.

$u = \dfrac{-11 \pm 5}{2 \bullet 4}$ Multiply in the denominator.

$u = \dfrac{-11 \pm 5}{8}$ Separate numerator.

$u = \dfrac{-11 + 5}{8}$ or $u = \dfrac{-11 - 5}{8}$ Add numerators and reduce fraction if possible.

$u = \dfrac{-6}{8} = \dfrac{-3}{4}$ or $u = \dfrac{-16}{8} = -2$

The solutions are $-\frac{3}{4}$ and -2. ∎

The last example will have two quadratic equations from section 8.6 that had no solutions. We will now see why they had previously no solutions.

EXAMPLE 3:

a)   Solve for x: $x^2 - x = -1$

Set equation to zero.
$x^2 - x + 1 = 0$
$a = 1$, $b = -1$, and $c = 1$.

Step 1: Place a, b and c into quadratic formula. $x = \dfrac{-(-1) \pm \sqrt{(-1)^2 - 4 \bullet 1 \bullet 1}}{2 \bullet 1}$

Step 2: Simplify using order of operations. Take power first inside radical.

$x = \dfrac{-(-1) \pm \sqrt{1 - 4 \bullet 1 \bullet 1}}{2 \bullet 1}$. Multiply inside radical.

$x = \dfrac{-(-1) \pm \sqrt{1 - 4}}{2 \bullet 1}$ Subtract inside radical.

$x = \dfrac{-(-1) \pm \sqrt{-3}}{2 \bullet 1}$ Take square root.

$x = \dfrac{-(-1) \pm i\sqrt{3}}{2 \bullet 1}$ Multiply in the denominator and numerator.

$x = \dfrac{1 \pm i\sqrt{3}}{2}$ Separate numerator.

$x = \dfrac{1 + i\sqrt{3}}{2}$ or $x = \dfrac{1 - i\sqrt{3}}{2}$

The solutions are $\dfrac{1 + i\sqrt{3}}{2}$ and $\dfrac{1 - i\sqrt{3}}{2}$.

b)   Solve for y: $y^2 + 9 = 0$

$a = 1$, $b = 0$, and $c = 9$.

Step 1: Place a, b and c into quadratic formula. $y = \dfrac{0 \pm \sqrt{0^2 - 4 \bullet 1 \bullet 9}}{2 \bullet 1}$

Step 2: Simplify using order of operations. Take power first inside radical.

$y = \dfrac{0 \pm \sqrt{0 - 4 \bullet 1 \bullet 9}}{2 \bullet 1}$. Multiply inside radical.

$y = \dfrac{0 \pm \sqrt{0 - 36}}{2 \bullet 1}$ Subtract inside radical.

$y = \dfrac{0 \pm \sqrt{-36}}{2 \bullet 1}$ Take square root.

$y = \dfrac{0 \pm 6i}{2 \bullet 1}$ Multiply in the denominator.

$y = \dfrac{0 \pm 6i}{2}$ Separate numerator.

$y = \dfrac{0 + 6i}{2}$ or $y = \dfrac{0 - 6i}{2}$ Add numerators and reduce fraction if possible.

$y = \dfrac{6i}{2} = 3i$ or $y = \dfrac{-6i}{2} = -3i$

The solutions are $3i$ and $-3i$.■

The two quadratic equations in example 3 have imaginary numbers in their solutions. An equation is said to have no real solutions when imaginary number is present in the solution. We were only concerned with equations back in section 8.6 that factored.

262

Equations which could not be factored had no solutions for us at that time. The quadratic formula finds real and imaginary solutions for every quadratic equation.

I will end this section with an example that has a radical in the solution.

EXAMPLE 4:

Solve for u: $2u^2 + 9u = -8$

Set equation to zero.
$2u^2 + 9u + 8 = 0$
$a = 2$, $b = 9$, and $c = 8$

Step 1: Place a, b and c into quadratic formula. $u = \dfrac{-9 \pm \sqrt{9^2 - 4 \bullet 2 \bullet 8}}{2 \bullet 2}$

Step 2: Simplify using order of operations. Take power first inside radical.

$u = \dfrac{-9 \pm \sqrt{81 - 4 \bullet 2 \bullet 8}}{2 \bullet 2}$. Multiply inside radical.

$u = \dfrac{-9 \pm \sqrt{81 - 64}}{2 \bullet 2}$ Subtract inside radical.

$u = \dfrac{-9 \pm \sqrt{17}}{2 \bullet 2}$ Multiply in the denominator.

$u = \dfrac{-9 \pm \sqrt{17}}{4}$ Separate numerator.

$u = \dfrac{-9 + \sqrt{17}}{4}$ or $u = \dfrac{-9 - \sqrt{17}}{4}$

The solutions are $\dfrac{-9 + \sqrt{17}}{4}$ and $u = \dfrac{-9 - \sqrt{17}}{4}$. ∎

The last example could have been considered having no solution back in section 8.6 because factoring would be impossible. It is clear that a real solution does exist. Section 8.6 only dealt with quadratic equations that were factorable. The quadratic formula deals with all quadratic equations; therefore, the quadratic formula generates true solution whether real or non real.

PROBLEMS:

Solve for the given variable using the quadratic formula.

1) $m^2 - 2m - 3 = 0$ 2) $n^2 - 7n + 6 = 0$

3) $o^2 + 8o + 15 = 0$

4) $p^2 + 4p - 21 = 0$

5) $q^2 - 4q = 12$

6) $r^2 + 5r = 14$

7) $2s^2 + 5s - 3 = 0$

8) $4t^2 - 24t + 35 = 0$

9) $6s^2 + 19s + 19 = 0$

10) $o^2 - 12o - 35 = 0$

11) $u^2 = -2u + 35$

12) $z^2 + 5z + 4 = 0$

13) $a^2 + 3a - 10 = 0$

14) $b^2 - 3b - 18 = 0$

15) $c^2 - 12c + 32 = 0$

16) $d^2 + 8d = -15$

17) $h^2 = 11h - 24$

18) $3g^2 + 7g + 2 = 0$

19) $6h^2 + 11h - 10 = 0$

20) $5j^2 + 2j = 3$

21) $16p^2 + 40p + 25 = 0$

22) $y^2 - 14y + 49 = 0$

23) $5q^2 + 13q = 6$

24) $r^2 - 2r = 0$

25) $s^2 = -8s$

26) $5t^2 - 15t = 0$

27) $u^2 - 25 = 0$

28) $v^2 = 81$

29) $2w^2 - 18 = 0$

30) $3y^2 + 24y + 45 = 0$

31) $6z^2 + 28z = 10$

32) $9y^2 - 12y + 4 = 0$

33) $4h^2 = 13h + 12$

34) $j^2 + 5j = 0$

35) $k^2 = 7k$

36) $16u^2 - 42u + 5 = 0$

37) $n^2 = 49$

38) $o^2 = 64$

39) $3p^2 - 75 = 0$

40) $4q^2 - 4q = 24$

41) $v^2 + 10v + 25 = 0$

42) $g^2 + 8g + 16 = 0$

43) $k^2 - 3k = 7$

44) $m^2 + 4m + 5 = 0$

SECTION 10.3
ZERO AND NEGATIVE EXPONENTS

We will revisit law #2 of exponents introduced in section 1.5 for this sections. Zero exponents will explain law 2b, and negative exponents will be known as law 2c. I will list all laws for law #2 of exponents at the end of this section.

EXAMPLE 1:

$$\frac{f^8}{f^8} = f^{8-8} = f^0 = 1$$

Notice that law 2a was used to get the zero exponent. Law 2b skips the subtraction when the exponents are the same.

Law 2a of exponents is $x^m \div x^n = x^{m-n}$ for $m > n$. Law 2b of exponents is $x^m \div x^n = 1$ for $m = n$. ■

Law 2c of exponents shows how to deal with exponents when $m < n$.

EXAMPLE 2:

$$\frac{f^8}{f^{10}} = f^{-2} = \frac{1}{f^2}$$

Using law 2a of exponents leaves a negative exponent. Negative exponents are not allowed law 2c shows how to change the exponent. Law 2c of exponents is

$$x^m \div x^n = \frac{1}{x^{n-m}} \text{ for } m < n. ■$$

A simple of way of stating law 2c of exponents is just subtract the smaller exponent from the larger one and leave the new exponent in the denominator.

I will give a small poem before showing the last example. I use to tell my students welcome to Poetry 101 before introducing negative exponents. I use a small rhyme to remind students on what to do when negative exponents are present. The rhyme is cross the line change the sign. The exponent of the variable changes its sign when the variable is moved from the numerator to the denominator or vice versa. Remember when there is no denominator present there is always a 1 in the denominator. f^2 crossed the line so that the exponent had to change signs.

EXAMPLE 3:

a) $\quad a^{-7} = \dfrac{1}{a^7}$

b) $\quad \dfrac{1}{b^{-3}} = b^3$

c) $\quad \dfrac{c^{-2}}{d^{-5}} = \dfrac{d^5}{c^2}$

LAW #2 OF EXPONENTS

a) $\quad x^m \div x^n = x^{m-n}$ for $m > n$

b) $\quad x^m \div x^n = \mathrm{x}^0 = 1$ for $m = n$

c) $\quad x^m \div x^n = \dfrac{1}{x^{n-m}}$ for $m < n$.

PROBLEMS:

Simplify.

1) $\quad f^4 \div f^4$

2) $\quad g^5 \div g^5$

3) $\quad h^3 \div h^4$

4) $\quad j^2 \div j^5$

5) $\quad 2^{-3}$

6) $\quad w^{-9}$

7) $\quad 3x^{-10}$

8) $\quad \dfrac{1}{3^{-3}}$

9) $\quad \dfrac{1}{\mathrm{y}^{-5}}$

10) $\quad \dfrac{z^{-6}}{a^{-4}}$

11) $\quad \dfrac{k^9}{l^{-1}}$

12) $\quad \dfrac{2^{-6}}{2^{-7}}$

13) $\quad \dfrac{6m^{-2}}{8n^{-7}}$

14) $\quad \dfrac{p^{-2}}{9q^{-1}}$

15) $\quad \dfrac{r^{-2}}{8s}$

SECTION 10.4
SCIENTIFIC NOTATION

Scientific notation is an alternative to writing large numbers or extremely small numbers. 1,000,000 is considered a large number and .000000001 is considered a small number.

The powers of 10 are required for scientific notation. $10^1 = 10$, $10^2 = 100$, $10^3 = 1000$, $10^4 = 10000$, ect. The exponent represents the number of zeros that come after the 1 in powers of 10. The powers of 10 also can be used for decimal numbers. $10^{-1} = .1$, $10^{-2} = .01$, $10^{-3} = .001$, $10^{-4} = .0001$, ect. The negative exponent represents the place value where the 1 is placed.

1×10^6 would be scientific notation for 1,000,000. 1×10^{-6} would be scientific notation for .000001. The main requirement for scientific notation is that the number before the multiplication by the power of 10 must be between 0 and 10.

Converting an integer or decimal number into scientific notation is a matter of sliding the decimal point. The decimal point is always at the end of an integer. The decimal point moves to the left when converting integers to scientific notation. The decimal point moves to the right when converting decimals to scientific notation.

EXAMPLE 1:

a)   Convert 2601 to scientific notation. Step 1: Move decimal point until the number becomes a number between 1 and 10.

2601 becomes 2.601. Step 2: Count the number of places the decimal point moved and use the count as a power of 10.

The decimal point moved 3 places to the left; therefore, the power of 10 is 10^3.

Step 3: Write the number in scientific notation. 2.601×10^3

b)   Convert .0075 to scientific notation. Step 1: Move decimal point until the number becomes a number between 1 and 10.

.0075 becomes 7.5. Step 2: Count the number of places the decimal point moved and use the count as a power of 10.

The decimal point moved 3 places to the right; therefore, the power of 10 is 10^{-3}.

Step 3: Write the number in scientific notation. 7.5×10^{-3} ∎

Now let's convert numbers written in scientific notation and convert them into integer or decimal form. We will use the opposite of the above procedure.

EXAMPLE 2:

a)   Convert 8.34×10^3 to a decimal number. Step 1: Move decimal point the number of places based on the power of 10's exponent.

8.34×10^3 becomes 8340. The decimal point was moved 3 places to the right.

b)   Convert 6.2×10^{-6} to a decimal number. Step 1: Move decimal point the number of places based on the power of 10's exponent.

6.2×10^{-6} becomes .0000062. The decimal point was moved 6 places to the left. ∎

SUMMARY

Integers and decimals → scientific notation

Integers – Move decimal point to the left.
Decimals – Move decimal point to the right.

Scientific notation → integers and decimals

Positive power of 10 – Move decimal point to the right.
Negative power of 10 – Move decimal point to the left.

PROBLEMS:

Convert to scientific notation.

1)	26.1	2)	126	3)	1263	4)	6.93
5)	.342	6)	.00352	7)	.6	8)	.000261126

Convert to decimal or integer form.

9)	6.11×10^2	10)	6.12×10^6	11)	3.83×10^7	12)	7.49×10^8
13)	5.72×10^{-6}	14)	6.112×10^{-6}	15)	1.263×10^{-8}	16)	4×10^{-3}

SECTION 10.5
FRACTIONAL EXPONENTS

Roots can be represented by fractional exponents. $\sqrt{x} = x^{\frac{1}{2}}$. The denominator is the root. A root of a number can always be taken when there is a one in the numerator. Just take the root according to the denominator. I will mention fractional exponents that have numbers greater than one in the numerator after the first example.

EXAMPLE 1:

Take the root of the given numbers.

a) $4^{\frac{1}{2}} = 2$ Just take the square root of 4.

b) $2985984^{\frac{1}{6}} = 12$ Just take the sixth root of 2985984. ∎

Fractional exponents that have numerators greater than one have a power and a root. The numerator represents the power and the denominator represents the root. $x^{\frac{3}{2}} = \left(\sqrt{x}\right)^3$ is an example how to rewrite the fractional exponent in radical form.

EXAMPLE 2:

a)   Rewrite in radical form. $a^{\frac{2}{5}} = \left(\sqrt[5]{a}\right)^2$ The denominator becomes the root and the numerator becomes the power.

b)   Rewrite in exponential form. $\left(\sqrt[7]{b}\right)^5 = b^{\frac{5}{7}}$ The power becomes the numerator and the root becomes the denominator. ∎

Now let's see how to calculate using fractional exponents.

EXAMPLE 3:

Simplify.

a) $4^{\frac{5}{2}}$ Step 1: Rewrite in radical form.
       $\left(\sqrt{4}\right)^5$ Step 2: Take the root.
       2^5 Step 3: Take the power.
       32.

b) $\quad 64^{\frac{7}{6}}$ Step 1: Rewrite in radical form.

$\left(\sqrt[6]{64}\right)^7$ Step 2: Take the root.

2^7 Step 3: Take the power.

128 ∎

Notice that Aunt Sally's rules are still in effect. The work was done inside the parentheses was done first.

Let us now look at how negative fractions exponents work. This discussion will be limited to what was studied before in section 10.3. Any further discussion would be done in an Intermediate Algebra text.

EXAMPLE 4:

Simplify.

a) $\quad 4^{-\frac{5}{2}}$ Step 1: Change the negative exponent to a positive exponent using the method from section 10.3.

$\dfrac{1}{4^{\frac{5}{2}}}$ Step 2: Rewrite in radical form.

$\dfrac{1}{\left(\sqrt{4}\right)^5}$ Step 3: Take the root.

$\dfrac{1}{2^5} 2^5$ Step 4: Take the power.

$\dfrac{1}{32}$.

b) $\quad 64^{-\frac{7}{6}}$ Step 1: Change the negative exponent to a positive exponent using the method from section 10.3.

$\dfrac{1}{64^{\frac{7}{6}}}$ Step 2: Rewrite in radical form.

$\dfrac{1}{\left(\sqrt[6]{64}\right)^7}$ Step 3: Take the root.

$\dfrac{1}{2^7}$ Step 4: Take the power.

$\dfrac{1}{128}$ ∎

PROBLEMS:

Take the root of the given numbers.

1) $64^{\frac{1}{2}}$ 2) $729^{\frac{1}{6}}$ 3) $1728^{\frac{1}{3}}$ 4) $2176782336^{\frac{1}{12}}$

Rewrite in radical form.

5) $a^{\frac{2}{3}}$ 6) $b^{\frac{2}{5}}$ 7) $d^{\frac{6}{31}}$ 8) $f^{\frac{3}{2}}$

Rewrite in exponential form.

9) $\left(\sqrt{g}\right)^5$ 10) $\left(\sqrt[11]{h}\right)^6$ 11) $\left(\sqrt[8]{j}\right)^7$ 12) $\left(\sqrt[3]{k}\right)^8$

Simplify

13) $25^{\frac{5}{2}}$ 14) $46656^{\frac{5}{6}}$ 15) $16^{\frac{5}{4}}$ 16) $64^{\frac{4}{3}}$

17) $4^{\frac{5}{2}}$ 18) $140608^{\frac{5}{3}}$ 19) $160000^{\frac{7}{4}}$ 20) $7962624^{\frac{3}{5}}$

SECTION 10.6
RAISING A POWER TO A POWER

The last law of exponents states the procedure to use when a power is raised to a power. $(x^2)^3$ is read x-squared raised to the third power. The third law of exponents states that when a power is raised to a power, the exponents are multiplied. $(x^2)^3 = x^{2 \cdot 3} = x^6$.

EXAMPLE 1:

Simplify $(b^3)^5$

Using the third law of exponents $(b^3)^5 = b^{15}$. ■

Now let us see how the third law is applied when coefficients are involved.

EXAMPLE 2:

Simplify $(3h^2)^6$

The first thing to do is distribute the exponent that is outside of the parentheses to everything inside the parentheses.

$(3h^2)^6 = (3^6)h^{2 \cdot 6}$

Now finish the expression.

$(3h^2)^6 = (3^6)h^{2 \cdot 6} = 729h^{12}$

The 3 and h^2 were raised to the sixth power. ■

The last example will compare the order of operations with the third law of exponents.

EXAMPLE 3:

a) Simplify $(5^2)^5$

The work inside parentheses is done first.

$(5^2)^5 = 25^5$

Now raise 25 to the fifth power.

$(5^2)^5 = 25^5 = 9765625.$

b) Simplify $(5^2)^5$ using the third law of exponents.

$(5^2)^5 = 5^{10} = 9765625$

The third law of exponents is an alternate way of raising powers of real numbers to a power. ∎

Here is a table with all of the laws of exponents used in this book.

LAW #1 OF EXPONENTS

$x^m x^n = x^{m+n}$ when the base is the same, just add the exponents

LAW #2 OF EXPONENTS

a) $\quad x^m \div x^n = x^{m-n}$ for $m > n$

b) $\quad x^m \div x^n = x^0 = 1$ for $m = n$

c) $\quad x^m \div x^n = \dfrac{1}{x^{n-m}}$ for $m < n$.

LAW #3 OF EXPONENTS

$(x^m)^n = x^{mn}$ just multiply the exponents when raising a power to a power.

PROBLEMS:

Simplify

1) $(b^2)^6$

2) $(F^{12})^2$

3) $(F^6)^2$

4) $(F^6)^7$

5) $(4c^3)^2$

6) $(16h^{11})^4$

7) $(9a^2)^5$

8) $(5b^6)^2$

9) $(2^2)^6$

10) $(6^3)^2$

11) $(2^6)^2$

12) $(4^4)^2$

13) $(a^2 b^6)^6$

14) $(c^{12} d^6)^2$

15) $(g^8 h^2)^3$

16) $(j^3 k^6)^5$

17) $(2a^6 b^6)^{12}$

18) $(6c^6 d^2)^6$

19) $(8g^6 h^3)^5$

20) $(3j^8 k^4)^9$

21) $(a^2 b^6 c^6)^{12}$

22) $(bc^2 d^6)^6$

23) $(p^8 q^8 r^2)^9$

24) $(r^3 s^3 t^5)^6$

25) $(8a^6 b^6 c^2)^2$

26) $(28b^6 c^{12} d^{16})^2$

27) $(19p^6 q^3 r^8)^3$

28) $(34r^6 s^9 t^5)^2$

BIBLIOGRAPHY

Aufmann, Richard N. and Barker, Vernon C. *Basic College Mathematics: An Applied Approach*. 5th Edition. Houghton Mifflin, 1995.

Beiser, Arthur. *Physics*. Fifth Edition. Addison-Wesley, 1991.

Brumbaugh, Doug. *Scratch Your Brain Where It Itches Book D-1: Algebra Games, Tricks, and Quick Activities*. Critical Thinking Press & Software, 1994.

Harnadek, Anita. *Critical Thinking Activities for Mathematics Book 4*. Critical Thinking Press & Software, 1991.

Johnson, L. Murphy and Steffensen, Arnold R. *Elementary Algebra*. Second Edition. Harper Collins, 1989.

Lial, Margaret L. and Miller, Charles D. *Beginning Algebra*. 5th ed. Harper Collins, 1988.

Millington, T. Alaric and Millington, William. *Dictionary Of Mathematics*. Perennial Library, 1966.

Pulsinelli, Linda and Hooper, Patricia. *Introductory Algebra: An Interactive Approach*. Second Edition. Macmillian, 1987.

Saunders, Hal. *When Are We Ever Gonna Have To Use This*? Updated Third Edition. Dale Seymour Publications, 1988.

Streeter, James et al. *Beginning Algebra*. Third Edition Form A. McGraw Hill, 1993.

Streeter, James et al. *Beginning Algebra*. Fourth Edition. WCB/McGraw Hill, 1998.

Zuckerman, Martin M. *Intermediate Algebra: A Straightforward Approach*. Alternate edition. Wiley, 1981.

ANSWERS

SECTION 1.1

1) $d + f.$

2) $p + 18.$

3) $j - k.$

4) $a - 2.$

5) $d - 3.$

6) $fh.$

7) $6g.$

8) $8jk.$

9) $2(c + d).$

10) $(b + c)(b - c).$

11) $c(c - 4).$

12) $a \div 2.$

13) $\dfrac{c + d}{2}.$

14) $\dfrac{d - f}{5}.$

15) $\dfrac{b + 5}{b - 5}.$

16) $\dfrac{f^2 + k}{29}.$

17) $b^2 - c.$

18) $a^3 + b.$

SECTION 1.2

1) 2^4

2) 3^5

3) j^5

4) $5d^3$

5) $f^4 h^2$

6) $5k^4 l^3$

7) $7sv^5$

8) 49

9) 77

10) 13

11) 2

12) 37

13) 624

14) 32

15) 576

16) 29

17) 1035

18) 141

19) 546

20) 1372

21) 3249

22) 234

23) 85

24) $(5 + 3) \bullet 2 - 4 + 1 = 13$

25) $4^2 - 2 + (3 \bullet 3) = 23$

SECTION 1.3

1) Commutative-addition

2) Associative-multiplication

3) Commutative-multiplication

4) Associative-addition

5) $3 + (4 + 5) = (3 + 4) + 5$
$3 + 9 = 7 + 5$
$12 = 12$

6) $2 \bullet (3 \bullet 6) = (2 \bullet 3) \bullet 6$
$2 \bullet 18 = 6 \bullet 6$
$36 = 36$

7) $2 \bullet (4 + 5) = 2 \bullet (4 + 5)$
$8 + 10 = 2 \bullet 9$
$18 = 18$

8) $3 \bullet (5 - 2) = 3 \bullet (5 - 2)$
$15 - 6 = 3 \bullet 3$
$9 = 9$

9) Answers may vary. One possible answer: $3 - 3 = 3 - 3$.

10) Answers may vary. One possible answer: $4 \div 4 = 4 \div 4$.

SECTION 1.4

1) $8c$

2) $15g^3$

3) $17mno$

4) $2k^2$

5) 0

6) p^2

7) $10r^2 s$

8) $24a^2$

9) $2b + 3c$

10) $3g + f$

SECTION 1.5

1) a^6

2) 3^4

3) d^6

4) g^{16}

5) $k^7 l^3$

6) c^{10}

7) a^{10}

8) $o^{15} p^{16}$

9) $6c^9$

10) $4a^9$

11) $30c^6 d^3$

12) $18b^5 c^3$

13) $12b^{15}$

14) $60d^{10} g^4$

15) $60b^{19}$

16) $378b^9 c^8$

17) b

18) g^{14}

19) c^{10}

20) kl

21) $2l$

22) $6b$

23) $2a^2 b$

24) $3b^2 g$

SECTION 1.6

1) 562

2) 505

3) 27

4) 4200

5) 100

6) 196

7) 9

8) 39

9) 1000

10) 133

11) 216

12) 70

13) 18

14) 6400

15) 160

16) 1240

17)   52   18)   218

19)   250 miles   20)   6¼ miles

SECTION 2.1

1)   444 or + 444   2)   -19

3)   -217   4)   274 or + 274

5)   4 or + 4   6)   -10

7)   10   8)   85

9)   12   10) 39

11) 0.17   12) 0.2

13) 1.3   14) 0.8

15) $\dfrac{6}{11}$ 16) $\dfrac{1}{10}$

17) $\dfrac{3}{11}$ 18) $\dfrac{7}{13}$

SECTION 2.2

1)   6   2)   23

3)   -9   4)   1

5)   -18   6)   - 2

7)   15   8)   0

9)   -10   10) 0

11) -2   12) -2.3

13) 0.2   14) 0.6

15) -3   16) 9

17) 12   18) 0

19)   1   20)   14

21)   6   22)   6

23) $\dfrac{10}{11}$ 24) $-\dfrac{7}{10}$

25) $\dfrac{1}{11}$ 26) $-\dfrac{8}{13}$

SECTION 2.3

1)   54   2)   -100

3)   -14   4)   -2.62

5)   6   6)   60

7)   33.28   8)   0

9)   0   10)   0

11) $-\dfrac{24}{11}$ 12) $-\dfrac{3}{5}$

13) $\dfrac{6}{11}$ 14)   0

15)   1   16)   -1

17)   24   18)   -72

19)   -72   20)   40

21)   0   22)   16

23)   42   24)   6

25)   9   26)   -6

27)   -1   28)   6

29)   10   30)   -7.5

| 31) | -2 | 32) | 11 |

SECTION 2.4

1)	7	2)	30
3)	-7	4)	-33
5)	0	6)	2
7)	undefined	8)	-73
9)	2	10)	-2.04
11)	2	12)	29
13)	15	14)	1

SECTION 2.5

1)	-33	2)	-8
3)	-12	4)	20
5)	-17	6)	156
7)	18	8)	-42
9)	72	10)	-14
11)	33	12)	9
13)	33	14)	279

SECTION 3.1

1)	$b = 56$	2)	$a = 10$
3)	$h = -22$	4)	$a = -6$
5)	$f = 5$	6)	$u = -29$
7)	$p = \frac{1}{2}$	8)	$m = \frac{2}{3}$
9)	$a = 1$	10)	$f = -1$

11) $j = 15$	12) $j = 5$
13) $q = 18$	14) $a = 15$
15) $a = -34$	16) $c = 52$
17) $x = 18$	18) $x = 15$
19) $m = 67$	20) $x = 22$

SECTION 3.2

1) $a = 14$	2) $a = 8$
3) $g = -7$	4) $a = -22$
5) $a = -5$	6) $n = 4$
7) $a = 7$	8) $a = 12$
9) $b = 63$	10) $s = 32$
11) $f = -20$	12) $p = -55$
13) $a = 2$	14) $c = -1$
15) $a = -2$	16) $f = 5$

SECTION 3.3

1) $a = 2$	2) $f = 1$
3) $c = 3$	4) $n = 2$
5) $a = -19$	6) $o = 4$
7) $n = -1$	8) $c = 2$
9) $b = -17$	10) $a = -5$
11) $a = -6$	12) $b = -4$
13) $i = -5$	14) $j = 5$
15) $a = -6$	16) $c = -16$

17) $s = 4$

18) $a = 5$

19) $z = -5/3$

20) $z = 1$

SECTION 3.4

1) $h = V/B$

2) $W = A/L$

3) $b = (7 - 2a)/3$

4) $b = (5 - 7c)/6$

5) $d = (q - af)/a$

6) $i = (w - bh)/b$

SECTION 3.5

1) 13

2) 10

3) 8, 9

4) 57,58,59

5) 12,14

6) 21,23

7) 9,12

8) 0,5

9) loser = 499 votes
winner = 630 votes

10) dryer = $173
washer = $193

11) dimes = 6
quarters = 8

12) Veronica is 7 years old
Betty is 12 years old

13) Juan is 14 years old
Maria is 18 years old

14) John is 19 years old
Jim is 27 years old

15) Frank = $1302.50
Betty = $1427.50

16) Martha weighed 5 pounds
Frank weighed 7 pounds

17) $484

18) 2 meters

19) $0.40

20) 48 points

21) 613 gallons

22) 71 push ups

SECTION 3.6

1) length = 33 inches
width = 16 inches

2) length = 10 centimeters
width = 4 centimeters

3) length = 21 inches
    width = 2 inches

4) length = 16 inches
    width = 10 inches

5) 63mph to the meeting
    52mph from the meeting

6) 5mph to the country
    10mph from the country

7) 11:00 AM

8) 4 hours

SECTION 3.7

1) <

2) >

3) <

4) >

5)
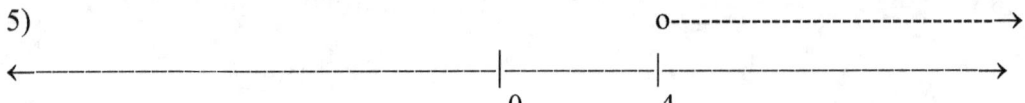

6)

7)

8)
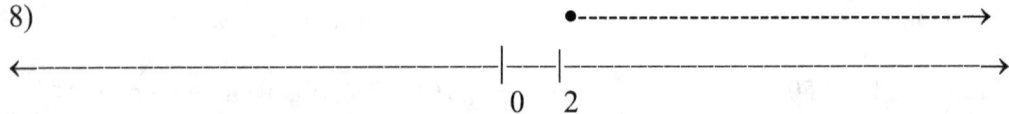

9)
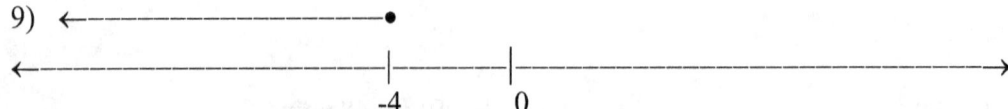

10)

11)
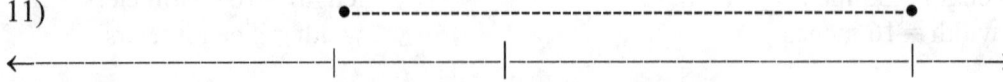

11)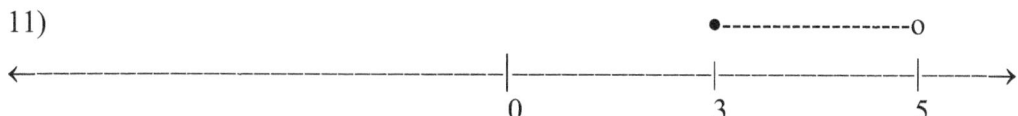

SECTION 3.8

1) $x < 5$ 2) $x \geq 13$

3) $x < 9$ 4) $x \geq 7$

5) $x < -5$ 6) $x \leq 3$

7) $x > -3$ 8) $x \leq -2$

9) $x \leq -3$ 10) $x < 3$

11) $x \geq 11$

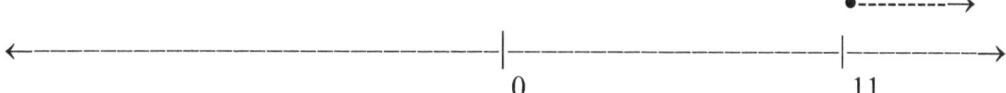

12) $x < -6$

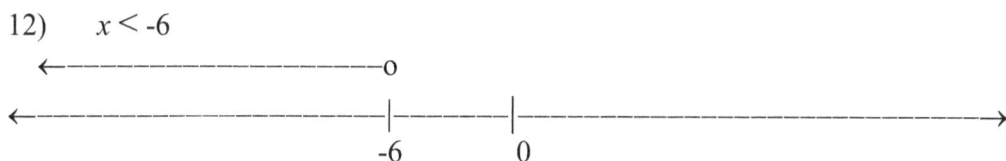

13) $x \leq -3$

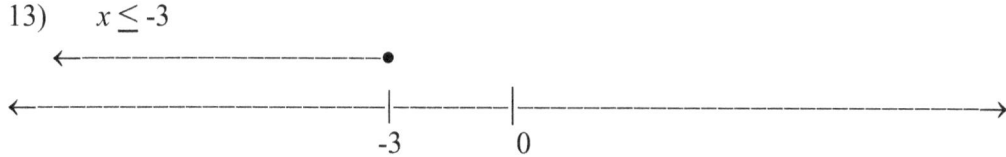

14) $x < 9$

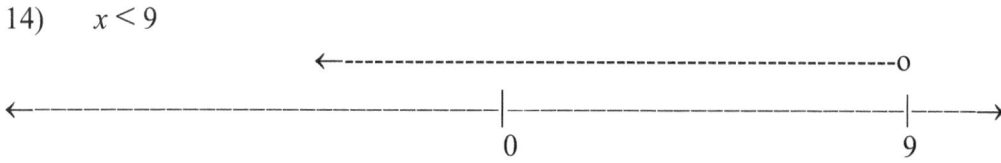

SECTION 4.1

1) $y = 9$

2) $y = 6$

3) $x = 1$

4) $x = 21$

5) $x = 6$

6) $y = 11$

7) $y = 1$

8) $x = 2$

9) (2,6), (-3,-24)

10) (12,0)

11) (1,-1), (2,3)

12) (13,-1)

13) (3,-3)

14) (9,-1)

15) (4,0), (8,3)

16) (3,4), (12,-2)

17) Answers may vary. Possible answers: (0,-6), (2,0), (1,-3), (-1,-9)

18) Answers may vary. Possible answers: (0,-14),(2,0), (1,-7), (-1,-21)

19) Answers may vary. Possible answers: (0,2), (18,0), (9,1), (-9,3)

20) Answers may vary. Possible answers: (0,-9), (2,0), (4,9), (-2,-18)

SECTION 4.2

1) Quadrant 1

2) Quadrant 4

3) Quadrant 2

4) Quadrant 3

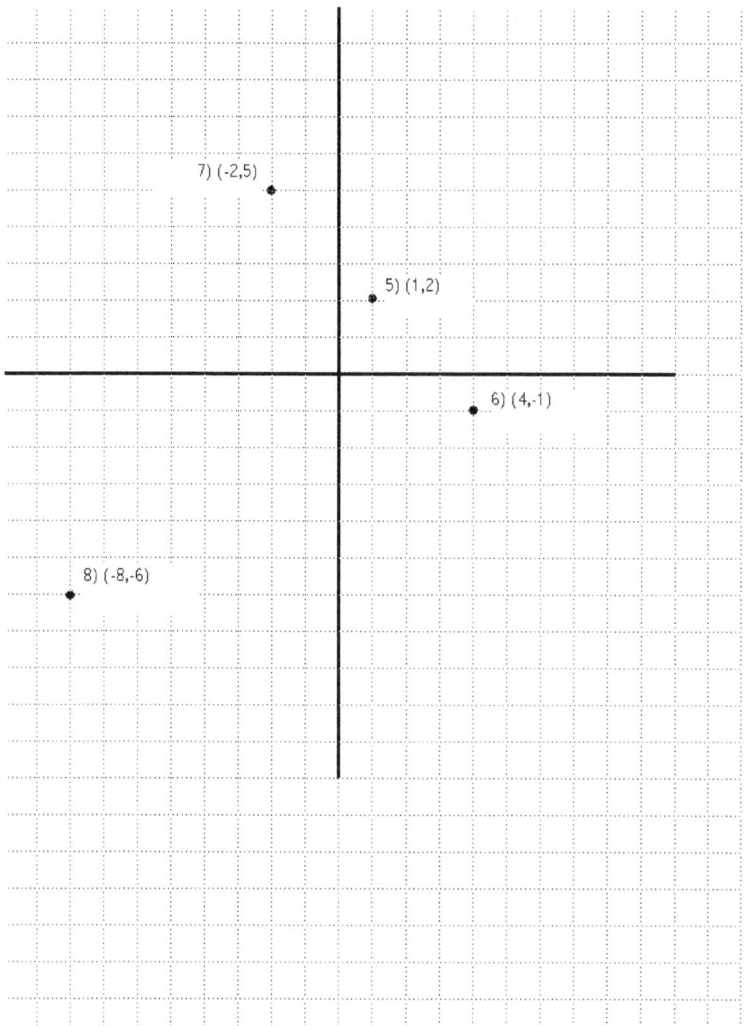

9)   (-2,-1)   10)   (-5,8)

11)   (4,-4)   12)   (3,2)

13)   (13/2, 25/2)   14)   (15/2, 4)

15)   (7/2, 33/2)   16)   (7/2, 5)

SECTION 4.3

1)

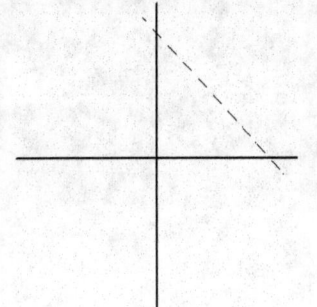

2)

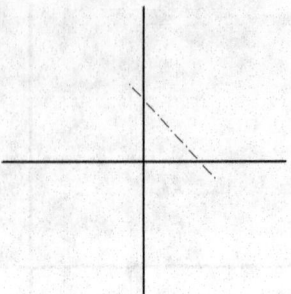

3)

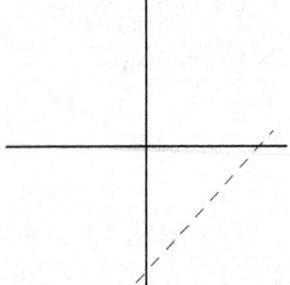

4)

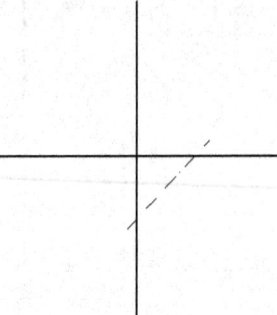

5)

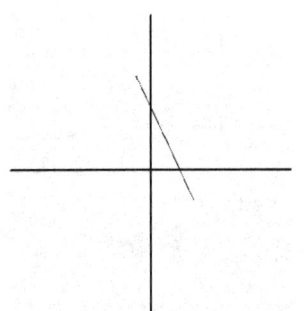

6)

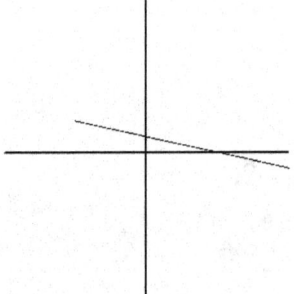

7)

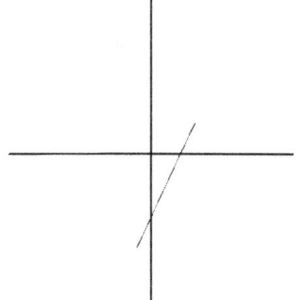

8)

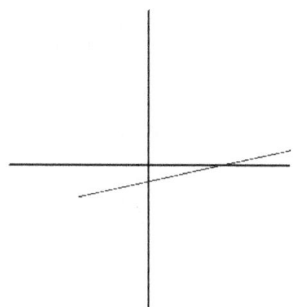

9)

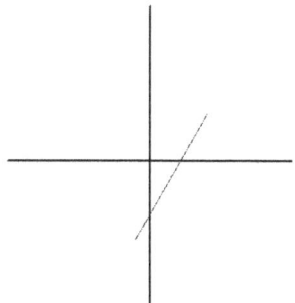

10)

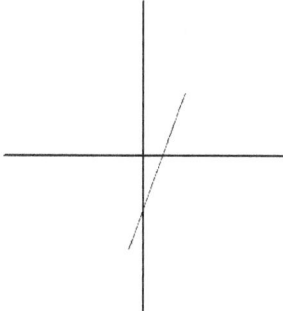

11)

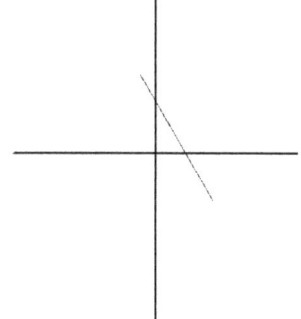

12)

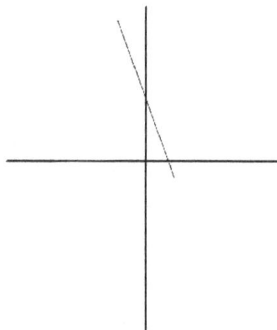

13)

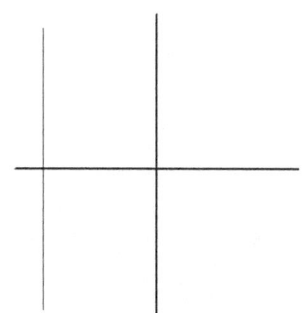

14)

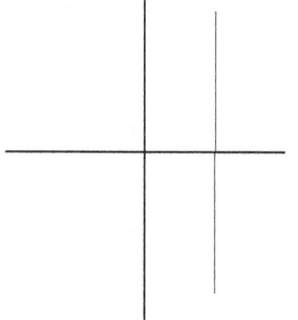

15)

16)

17)

18)

19)

20)

SECTION 4.4

1)

2)

3)

4)

5)

6)

7)

8)

9)

10)

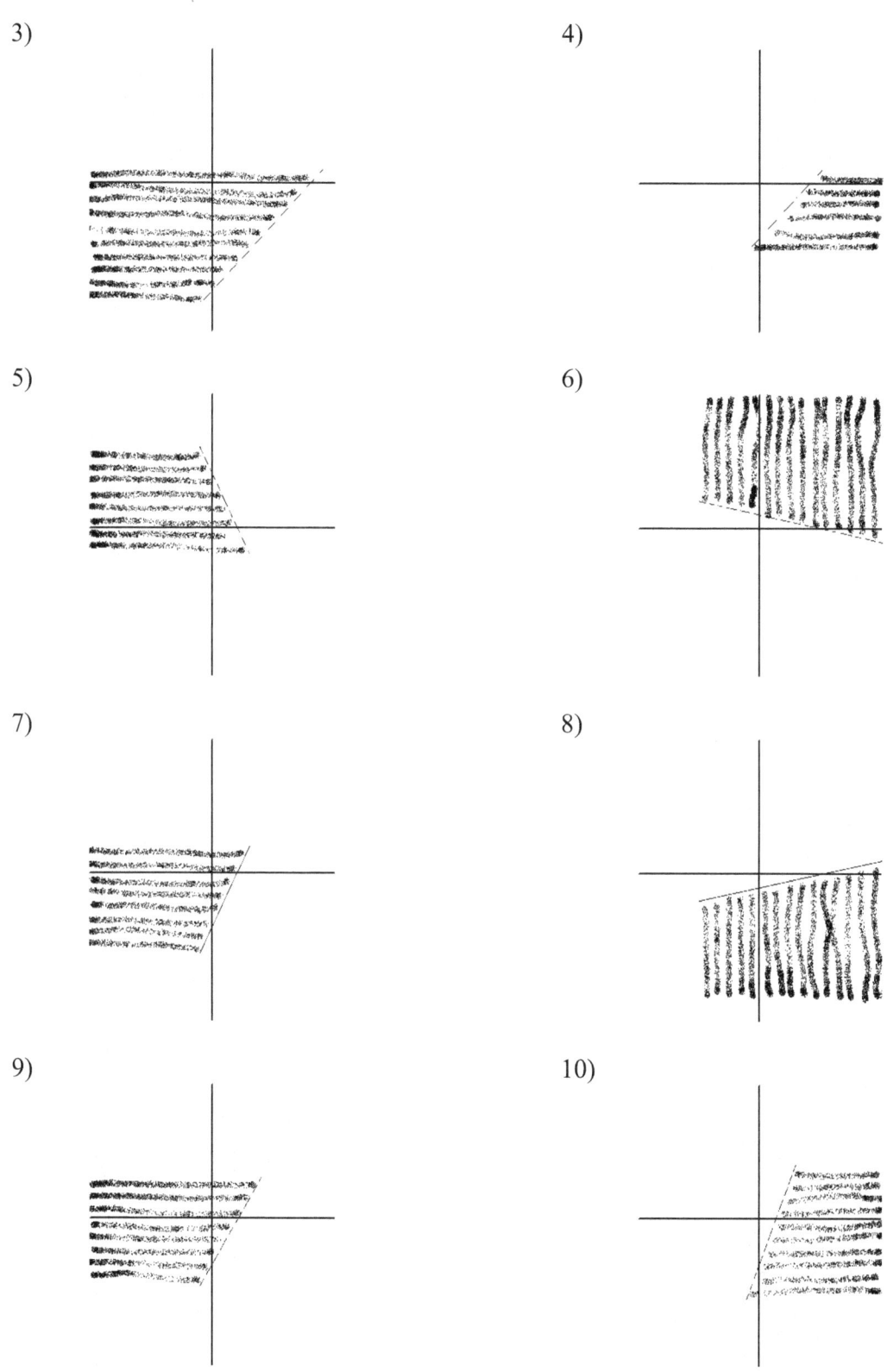

290

11)

12)

13)

14)

15)

16)

17)

18)

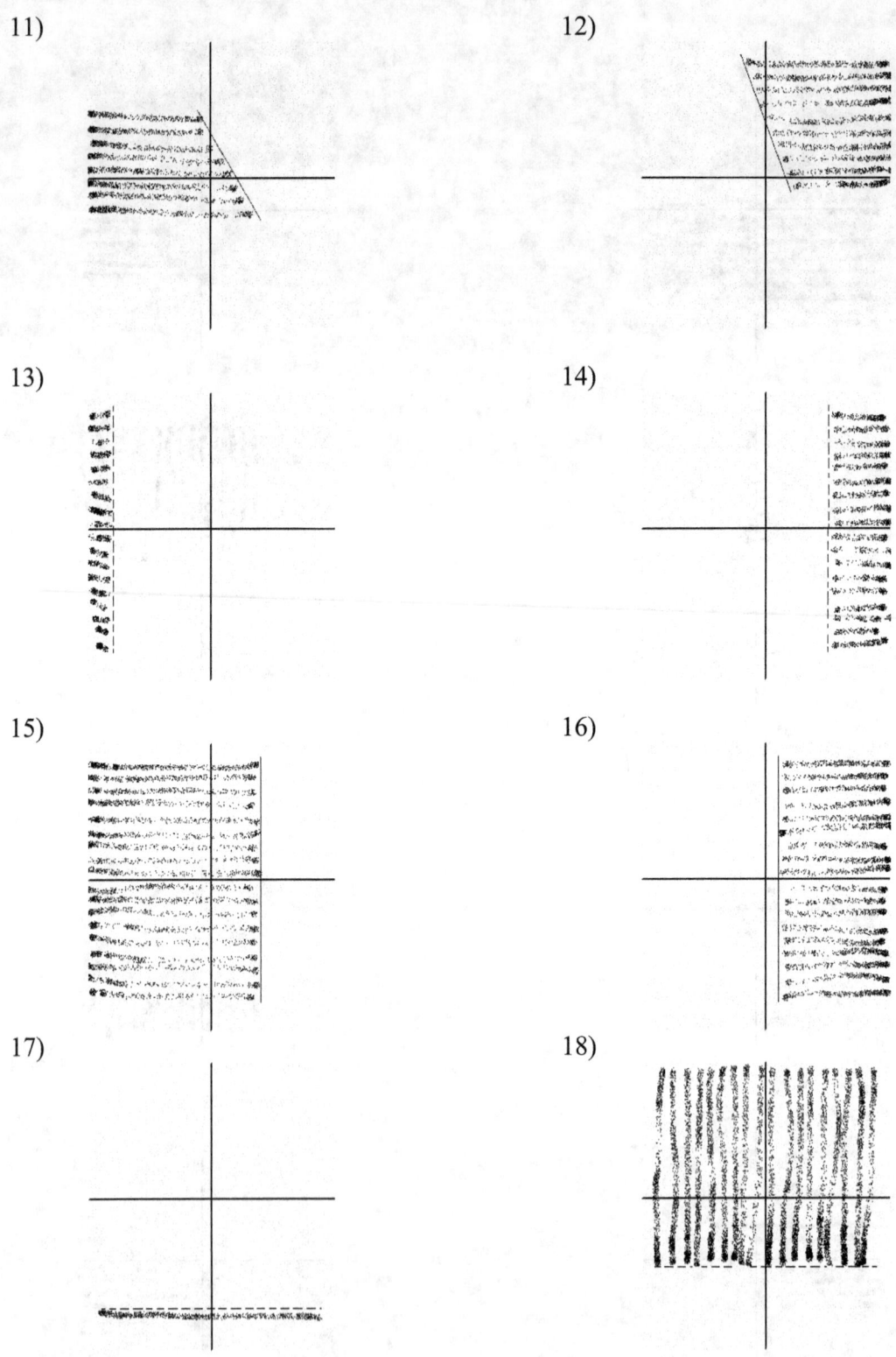

19)

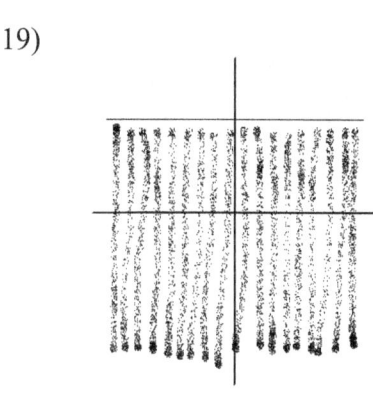

20)

SECTION 5.1

1)   (4, -2)

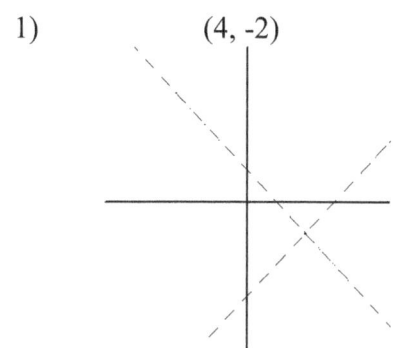

2)   (1, 7)

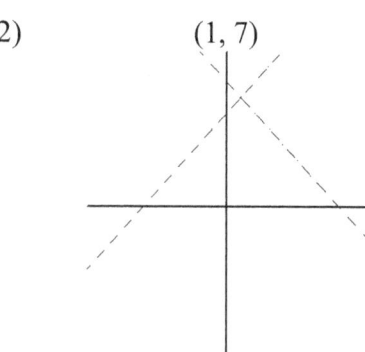

3)   (2, 3)

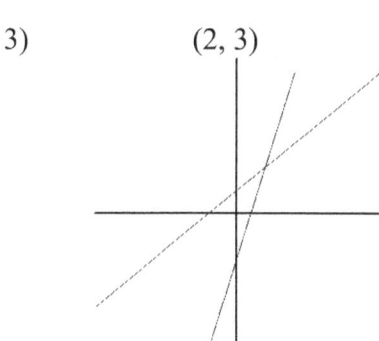

4)   (5, 2)

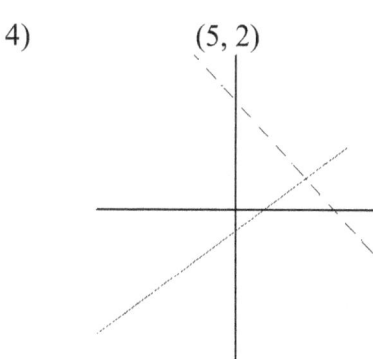

5)   (3, 1)

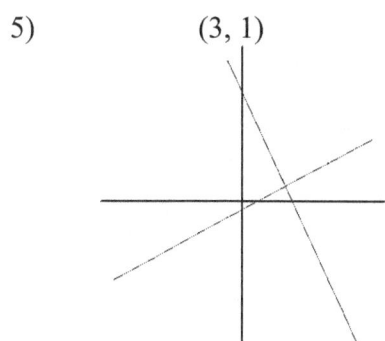

6)   (6, -3)

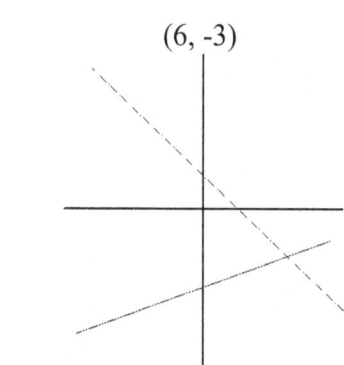

292

7)   no solution

8)   infinite solutions

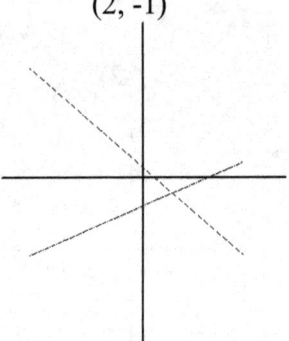

9)   no solution

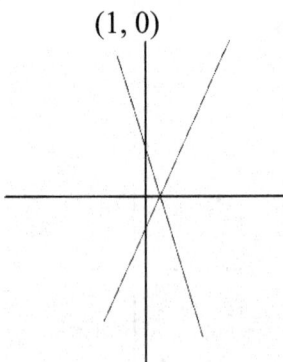

10)   (2, -1)

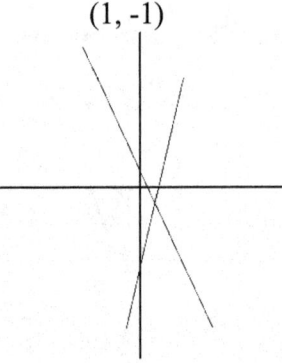

11)   (1, 0)

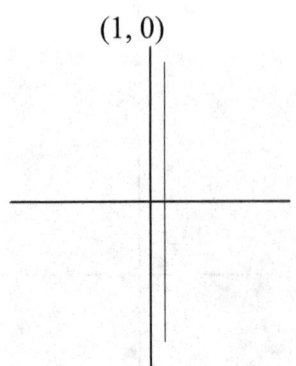

12)   (1, -1)

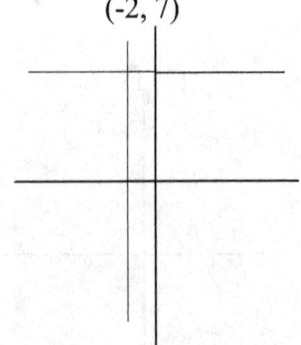

13)   (1, 0)

14)   (-2, 7)

15)   (-1, -8)

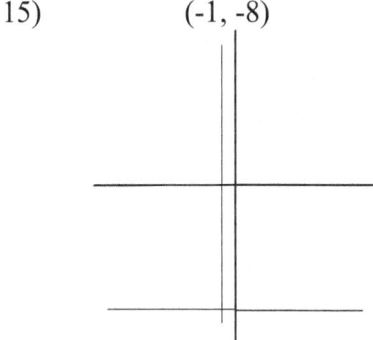

16)   (2, -1)

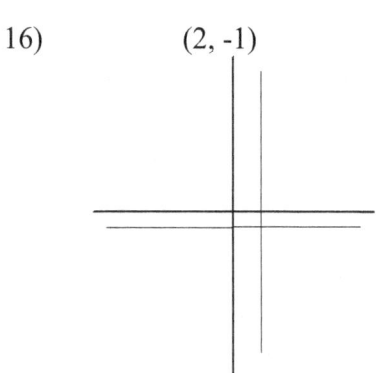

SECTION 5.2

1)   (4,-2)

2)   (1,7)

3)   (2,3)

4)   (5,2)

5)   (8,-6)

6)   (6,-3)

7)   inconsistent

8)   dependent

9)   inconsistent

10)   (-24,22)

11)   (1/2,7)

12)   (2/3,-1/3)

SECTION 5.3

1)   (4,-2)

2)   (1,7)

3)   (2,3)

4)   (5,2)

5)   (8,-6)

6)   (6,-3)

7)   inconsistent

8)   dependent

9)   inconsistent

10)   (3,21)

SECTION 5.4

1)   17 and 15

2)   5 dimes and 10 quarters

3)   2 nickels and 25 dimes

4)   22 tens and 63 twenties

5)   80 fives and 44 tens

6)   writing pads = $1.01
     clipboards = $1.35

7)   highlighter = $0.41
     marker = $1.05

8)   candy bar = 60¢
     pack of gum = 35¢

9)   apple = 49¢
     pear = 59¢

10) 15% solution = 52ml
     20% solution = 78ml

11) 20% solution = 420ml
     35% solution = 80ml

12) 30% solution = 50ml
     15% solution = 1000ml

13) $8500 at 7%
     $1500 at 5%

14) $4875 at 11%
     $1125 at 7%

15) $3000 at 8%
     $5000 at 12%

16) 300 tickets at $6
     400 tickets at $8

17) 7.5 pounds of peanuts
     22.5 pounds of walnuts

SECTION 6.1

1)   monomial

2)   binomial

3)   trinomial

4)   ordinary

5)   ordinary

6)   monomial

7)   polynomial

8)   not a polynomial

9)   not a polynomial

10) polynomial

11) third degree

12) seventh degree

13) second degree

14) zero degree

15) $-5x^3 + 10x^2 + x - 3$

16) $-11y^4 + 9y^3 + 6y + 12$

17) $-2y^5 + y^3 - y^2 - 5y + 1$

18) $5x^5 - 6x^4 + 2x^3 + x^2 + 11x + 1$

SECTION 6.2

1) $9f - 4$

2) $10c^2 - 24c$

3) $-3h^2 - 2h$

4) $9d^2 - 4d$

5) $-3d - 5g$

6) $-6f + 7h$

7) $7r^2 - 8r + 9$

8) $3a^6 + 2a^4 - 8a^2$

9) $2k - 5$

10) $-4j^2$

11) $-5n^2 + n$

12) $4m^2 - 3m - 3$

13) $4g^2 + 2g + 3$

14) $2a^3 + 5a^2 - 4a$

15) $-4b^2 + 2b - 1$

16) $d^2 + 4d - 7$

17) $8c^2 - 12c - 7$

18) $y^3 - y^2 - 5y$

19) $-5b^2 - 3b - 2$

20) $-9h^5 + 4h^4 - 3h^3 - 8h^2 + 2h$

21) $-7g^2 - 10g - 4$

22) $-3h + 4$

23) $5z^2 - 3z + 9$

24) $5a^2 - 2a + 4$

25) $d^2 + 7d - 11$

26) $9k^2 + 6k - 8$

27) $37h^2 - 3h + 13$

28) $-2j - 14$

SECTION 6.3

1) $a^2 + 8a + 12$

2) $g^2 - 9g + 14$

3) $c^2 + c - 12$

4) $d^2 - 5d - 14$

5) $k^2 - 16$

6) $h^2 - g^2$

7) $j^2 - 1.21$

8) $m^2 - \dfrac{4}{9}$

9) $14q^2 + 25qd + 6d^2$

10) $24e^2 + 42eg - 12g^2$

11) $140x^2 - 244xy + 99y^2$

12) $1.8k^2 + 8.85km - 11.7m^2$

13) $15cf + 18cg - 20hf - 24hg$

14) $32jn - 24np - 8oj + 6op$

15) $a^2 + a + \frac{1}{4}$

16) $g^2 - 18g + 81$

17) $k^2 - 0.1k - 1.56$

18) $a^2 - 8.3a + 17.22$

19) $63,063$

20) $41,832$

21) 428,770

22) 344,463

23) $a^3 - 11a^2 - 29a + 39$

24) $s^3 + 2s^2 - 9s + 20$

25) $12g^4 + 4g^3 - 32g^2 - 20g + 26$

26) $a^4 - 5a^3 - 12a^2 - 5a + 1$

27) $24a^3 - 32a^2 - 6a$

28) $16c^3 - 28c^2 + 10c$

SECTION 6.4

1) $3v^8 + 2v^3 + 4$

2) $m^2 - 3m$

3) $2s^5 + s^3 - 2s$

4) $13t^4 - t^2 - 1$

5) 200 R 25

6) 35 R 47

7) 271

8) 608

9) $b + 2$

10) $g + 7$

11) $t - 5$

12) $h + 1$

13) $x + 3$

14) $s + 1$

15) $a - 6$

16) $n + 1$

17) $d + 1$

18) $c + 7$

19) $m^2 + 2m + 2$

20) $h^3 - 1$

21) $2n + 3 + \dfrac{5}{3n - 5}$

22) $2v - 4 + \dfrac{3}{2v + 7}$

23) $d^2 - 2d + 3 + \dfrac{1}{2d + 1}$

24) $2m^2 - 2m + 1 - \dfrac{1}{4m + 1}$

25) $2v + 3 + \dfrac{4}{v - 3}$

26) $4d + 2 - \dfrac{6}{d + 6}$

27) $m^3 + m^2 + m + 1$

28) $v^2 + 2v + 4$

29) 9 through 20, 27 and 28. All problems can be written as fractions. The polynomials can be factored. The fractions can be reduced.

30)   21 through 26. All problems can be written as fractions. The polynomials cannot be factored. The fractions cannot be reduced.

SECTION 7.1

1)   The square root of 7.   2)   The cube root of 75.

3)   The fifth root of 59.   4)   The square root of 800.

5)   2   6)   4

7)   3   8)   30

9)   16   10)   14

11)   8   12)   23

13)   2.828   14)   4.198

15)   3.222   16)   22.361

17)   20.543   18)   14.213

19)   9.818   20)   13.266

SECTION 7.2

1) $\sqrt{30}$ 2) $\sqrt{8}$

3) $\sqrt{56}$ 4) $\sqrt{42}$

5) $\sqrt[3]{16}$ 6) $\sqrt[3]{45}$

7) $\sqrt[5]{27}$ 8) $\sqrt[4]{20}$

9) $\sqrt{3850}$ 10) $\sqrt{1134}$

SECTION 7.2

1) $2\sqrt{17}$ 2) $4\sqrt{3}$

3) $2\sqrt[3]{2}$ 4) $2\sqrt[3]{12}$

5) $\sqrt{10}$

6) $\sqrt{74}$

7) $\sqrt[3]{28}$

8) $\sqrt[3]{95}$

9) h^4

10) j^5

11) b^{27}

12) h^4

13) $p^4\sqrt{p}$

14) $r^8\sqrt{r}$

15) $c\sqrt[5]{c^3}$

16) $h^5\sqrt[4]{h^2}$

17) $5h^4\sqrt{3}$

18) $2c\sqrt{11c}$

19) $2o^3p\sqrt{21op}$

20) $4m^4n^4\sqrt{2n}$

21) $2w^2\sqrt[3]{w^2}$

22) $2d^2\sqrt[4]{5d^2}$

23) $2mn^2\sqrt[3]{7m^2}$

24) $2xyz\sqrt[5]{3x^3z}$

25) $3y^4\sqrt{y}$

26) $\sqrt{10a}$

27) $m^2\sqrt[5]{12m^2}$

28) $n^7\sqrt[4]{2}$

29) $6q\sqrt{2q}$

30) $m^4\sqrt{47m}$

31) $20x^2y^4\sqrt{2y}$

32) $10\sqrt{3a}$

33) $2g^4\sqrt[3]{50g}$

34) $2m^2\sqrt[4]{30m}$

35) $2mn^3\sqrt[3]{70mn}$

36) $2y^2\sqrt[5]{30x^3z}$

SECTION 7.4

1) $4\sqrt{2}$

2) $15\sqrt{3}$

3) $\sqrt{2a}$

4) $-3\sqrt{3x}$

5) $6\sqrt{7}$

6) $-3\sqrt{10}$

7) $5\sqrt{3} - \sqrt{2}$

8) $2\sqrt{7} + 7\sqrt{6}$

9) $18\sqrt{5x}$

10) $7\sqrt{3b}$

11) $20\sqrt{7y}$

12) $12\sqrt{5c}$

13) $-5\sqrt{10z}$

14) $\sqrt{7d}$

15) $12\sqrt{3}$

16) $10\sqrt{2}$

17) $-6b\sqrt{2}$

18) $4c^2\sqrt{5}$

19) $12y\sqrt{3y}$

20) $30d^2\sqrt{2d}$

21) $19\sqrt{11}$

22) $20\sqrt{13}$

23) $3g\sqrt{2g}$

24) $2h^3\sqrt{3h}$

25) $17\sqrt{5} - 3\sqrt{3}$

26) $25\sqrt{6} - 9\sqrt{7}$

27) $17\sqrt{2}$

28) $40\sqrt{3}$

29) $-20\sqrt{5}$

30) $14\sqrt{7} - 12\sqrt{11}$

SECTION 7.5

1) $2\sqrt{3}$

2) $2\sqrt{6}$

3) $5a\sqrt{2}$

4) $7b^2\sqrt{2}$

5) $2\sqrt[3]{6}$

6) $3c$

7) $4\sqrt{10}$

8) $160\sqrt{21}$

9) $224d\sqrt{2}$

10) $78h^8\sqrt{22}$

11) $4\sqrt[3]{12}$

12) $120g^3$

13) $2\sqrt{10} + 5\sqrt{2}$

14) $\sqrt{77} - 2\sqrt{21}$

15) $2\sqrt{143h} + 2\sqrt{33h}$

16) $3\sqrt{130j} + 3j\sqrt{182}$

17) $2 + \sqrt{10} + \sqrt{6} + \sqrt{15}$

18) $6 - 4\sqrt{3} + \sqrt{42} - 2\sqrt{14}$

19) -1

20) -21

21) 2

22) $\sqrt{2}$

23) $2\sqrt{2}$

24) 3

25) 4

26) 2

27) $4k$

28) $3m\sqrt{5}$

29) $2\sqrt{2}$

30) $3\sqrt{2}$

31) $2\sqrt{10}$

32) 27

33) $14n\sqrt{13}$

34) $5o\sqrt{5}$

SECTION 8.1

1) 24,12,8,6,4,3,2,1

2) 45,15,9,5,3,1

3) 508,254,127,4,2,1

4) 9250,4625,1850,925,724,370,250,185,125,74,50,37,25,10,5,2,1

5) $7(h + 9)$

6) $2(g - 10)$

7) $5(y + 11)$

8) $6(q - 112)$

9) $q(1 + r)$

10) $b(1 - a)$

11) $t(s + 1)$

12) $z(y - 1)$

13) $u(u^2 + u - 1)$

14) $n(n^3 + n^2 - n^1 + 1)$

15) $p(p^4 - p^3 + p^2 - p + 1)$

16) $k^2(k^4 - k^2 + 1)$

17) $2b(2b^3 + 3)$

18) $3d(d^2 - 5d + 11)$

19) $3pq(8p^2q - 11p + 32q^2 - 4)$

20) $2xy(30x^2y^4 + 3xy^3 - 4xy^5 + 1)$

SECTION 8.2

1) $(a + 7)(a - 7)$

2) $(b + 2)(b - 2)$

3) $(c + 1)(c - 1)$

4) $(d + 22)(d - 22)$

5) $(3h + 6)(3h - 6)$

6) $(4g + 5)(4g - 5)$

7) $(9h + 8)(9h - 8)$

8) $(100j + 11)(10j - 11)$

9) $(7m + 10n)(7m - 10n)$

10) $(9o + 12p)(9o - 12p)$

11) $(6q + 11r)(6q - 11r)$

12) $(8s + 13g)(8s - 13g)$

13) $(5 + u)(5 - u)$

14) $(7 + v^2)(7 - v^2)$

15) $(4 + w^3)(4 - w^3)$

16) $(6 + x)(6 - x)$

17) $2y(3y + 1)(3y - 1)$

18) $3z(2z + 5)(2z - 5)$

19) $3a^2(4 + 3a)(4 - 3a)$

20) $2b(5 + b)(5 - b)$

21) $(c^2 + 10)(c^2 - 10)$

22) $(dh + 13)(dh - 13)$

23) $(3 + eg)(3 - eg)$

24) $(h^2 + 15j)(h^2 - 15j)$

25) $2k(3k + m)(3k - m)$

26) $3no(2n + 5o)(2n - 5o)$

SECTION 8.3

1) $(p + 4)(p + 2)$

2) $(q + 5)(q + 3)$

3) $(r + 3)(r + 10)$

4) $(s + 3)(s + 11)$

5) $(u - 4)(u - 5)$

6) $(v - 4)(v - 7)$

7) $(w - 3)(w - 5)$

8) $(y - 7)(y - 7)$

9) $(z + 3)(z - 2)$

10) $(a - 3)(a + 5)$

11) $(b - 2)(b + 5)$

12) $(c + 11)(c - 5)$

13) $(d-9)(d+7)$ 14) $(h+4)(h-5)$

15) $(g+4)(g-10)$ 16) $(h-12)(h+2)$

17) not factorable   18) not factorable

19) not factorable   20) not factorable

21) $2(p+5)(p+3)$ 22) $8(u+7)(u+1)$

23) $9(b+5)(b-4)$ 24) $3(h-11)(h+9)$

25) $(j+5)(j+5)$ 26) $6(k+2)(k-10)$

27) $2m^4(m+3)(m-7)$ 28) Yes. $(y-7)^2$

29) Yes. $(j+5)^2$

30) Yes. Multiplication is commutative and the order of the factors does not matter.

SECTION 8.4

1) $(p+3)(p+1)$ 2) $(q-3)(q+1)$

3) $(r-7)(r-5)$ 4) $(s-3)(s+8)$

5) $(u-7)(u-2)$ 6) $(v+5)(v+5)$

7) $(w-9)(w+7)$ 8) $(y-7)(y-7)$

9) $(z-7)(z+11)$ 10) $(a+5)(a+1)$

11) $(b-5)(b-3)$ 12) $(c-3)(c-7)$

13) $(d+9)(d+3)$ 14) $(h-6)(h-5)$

15) $(g+4)(g+4)$ 16) $(h-8)(h+6)$

17) $(j-27)(j-3)$ 18) $(k-3)(k+11)$

19) $(m-1)(m+3)$ 20) $(o+7)(o+3)$

21) $2(p-5)(p+1)$ 22) $3(u-3)(u-1)$

23) $3(b-5)(b-1)$ 24) $2(h-7)(h+6)$

25) $(j - 6)(j + 12)$

26) $3(k - 8)(k + 4)$

27) $2 m^4 (m - 10)(m + 5)$

28) Yes. $(v + 5)^2$

29) Yes. $(y - 7)^2$

30) Yes. $(g + 4)^2$

SECTION 8.5

1) $(2p - 1)(p + 3)$

2) $(3q - 2)(2q - 5)$

3) $(3r + 1)(r - 5)$

4) Not factorable.

5) $(3u + 1)(2u - 3)$

6) $(3v + 2)(5v - 3)$

7) $(6w - 5)(w + 5)$

8) $(3y - 2)(3y - 2)$ or $(3y - 2)^2$

9) $(2z - 3)(6z + 5)$

10) $(3a - 2)(a + 3)$

11) $(b - 4)(8b + 5)$

12) $(2c + 1)(c + 1)$

13) $(d - 2)(5d + 2)$

14) $(9h - 5)(h + 1)$

15) $(2g - 3)(3g - 4)$

16) $(12h - 5)(3h + 1)$

17) $5(2j - 3)(2j + 1)$

18) $(4k + 1)(3k + 4)$

19) $(2m - 3)(4m + 3)$

20) $(4o - 5)(5o - 2)$

21) $(4p + 5)(4p + 5)$ or $(4p + 5)^2$

22) $(4u - 3)(3u + 5)$

23) $(8b - 9)(3b + 4)$

24) $6(h - 1)(4h + 1)$

25) $3(p + q)(3p + 7q)$

26) $2(3u - 4v)(2u + 3v)$

27) $(b + 3c)(4b + c)$

28) $(3j + k)(2j - 5k)$

29) $(k + 3)(2k + 1)$

30) $(3d - 5)(3d - 5)$ or $(3d - 5)^2$

31) $(2h - 3)(5h + 4)$

32) $(4g - 3)(2g + 5)$

33) $(9x - 2y)(9x - 2y)$ or $(9x - 2y)^2$

34) $(2x - 7y)(3x + 8y)$

35) $(h + 5)(3h + 1)$

36) $(5j - 2)(5j - 2)$ or $(5j - 2)^2$

37) $(2u - v)(u + 7v)$

38) $(3x - y)(x - 3y)$

SECTION 8.6

1)	-1, 3		2)	1, 6
3)	-3, -5		4)	-7, 3
5)	-2, 6		6)	-7, 2
7)	-3, 1/2		8)	5/2, 7/2
9)	-3/4, -2		10)	No solution
11)	7		12)	-4, -1
13)	-5, 2		14)	-3, 6
15)	4, 8		16)	-3, -5
17)	3, 8		18)	-1/3, -2
19)	-5/2, 2/3		20)	-1, 3/5
21)	-5/4		22)	No solution
23)	-3, 2/5		24)	0, 2
25)	0, -8		26)	0, 3
27)	-5, 5		28)	-9, 9
29)	-3, 3		30)	-5, -3
31)	-5, 1/3		32)	2/3
33)	-3/4, 4		34)	-5, 0
35)	0, 7		36)	-5, 0
37)	7, -7		38)	8, -8
39)	-5, 5		40)	-2, 3
41)	4, -5		42)	-4
43)	-5		44)	No solution

45) No solution   46) No solution

47) 2, -2   48) 2, -2

49) 1/8, 5/2   50) 5, -7

SECTION 8.7

1) $h = \dfrac{f}{d + g}$

2) $j = \dfrac{h}{f - g}$

3) $\dfrac{mj}{j + m} = k$

4) $\dfrac{C}{1 - s} = L$

5) $\dfrac{mk}{m - k} = j$

6) $y = \dfrac{p}{n - m}$

7) $q = \dfrac{p}{p + 2}$

8) $\dfrac{sv}{v - s} = t$

9) $\dfrac{st}{t - s} = v$

10) $\dfrac{C}{1 + s} = L$

11) $\dfrac{qf}{q - f} = p$

12) $\dfrac{pf}{p - f} = q$

SECTION 8.8

1) -1 or 4

2) -7 or 2

3) 6, 11 or -6, -11

4) 5, 12 or -5, -12

5) 8 and 15

6) 4 and 14

7) -9 or 8

8) -7 or 8

9) 5 ft. by 13 ft.

10) 10 cm by 14 cm

11) 7 ft by 10 ft

12) 10 cm by 15 cm

13) -9, -13 or 9, 13

14) -13, -4 or 4, 13

15) -3 and -10

16) 5 and 11

17)   11, 12 or -12, -11   18)   10 and 12

SECTION 9.1

1)   yes   2)   no

3)   yes   4)   no

5)   yes   6)   no

7)   0   8)   none

9)   −3 and 3   10)   3 and 4

11)   −1 and 2/3   12)   none

13)   −3   14)   5

SECTION 9.2

1) $\dfrac{3}{10}$ 2) $\dfrac{1}{25}$

3) $\dfrac{-1}{11}$ 4) $\dfrac{-2}{27}$

5) $\dfrac{3}{5}$ 6) $\dfrac{3}{7}$

7) $\dfrac{5}{m+2}$ 8) $\dfrac{5(d+2)}{d+6}$

9) $\dfrac{h-1}{h+3}$ 10) $\dfrac{2g+1}{g+1}$

11) $\dfrac{r-5}{r-2}$ 12) $\dfrac{q+1}{q+5}$

13) $\dfrac{3h+1}{4h+5}$ 14) $\dfrac{2g+13}{3(g+2)}$

15) $\dfrac{h+2}{h+8}$ 16) $\dfrac{j+5}{j+4}$

17) $\dfrac{-2}{5+k}$

18) $\dfrac{-3}{4+m}$

19) -1

20) -1

21) $\dfrac{1}{p-3}$

22) $-q-3$

23) $r+5$

24) $\dfrac{1}{s-6}$

25) $\dfrac{t-6}{t-1}$

26) $\dfrac{u-5}{u+4}$

27) $b+2$

28) $g+7$

29) $t-5$

30) no.

SECTION 9.3

1) $\dfrac{2}{65}$

2) $\dfrac{1}{84}$

3) $\dfrac{1}{6}$

4) $\dfrac{1}{10}$

5) $\dfrac{6}{q}$

6) 2

7) $\dfrac{2u-1}{2u-5}$

8) $\dfrac{s+3}{3s}$

9) $\dfrac{(t+2)(t+10)}{(t+5)(t-5)}$

10) $\dfrac{3r^2}{r+8}$

11) $\dfrac{5}{2}$

12) $\dfrac{55}{4}$

13) $\dfrac{50}{13}$

14) 2

15) $\dfrac{1}{6v}$

16) $\dfrac{w-3}{10w}$

17)   1

18) $\dfrac{y-2}{3y^2}$

19) $\dfrac{z-7}{z-3}$

20) $\dfrac{2(a+2)}{9}$

SECTION 9.4

1) $\dfrac{6}{17}$

2) $\dfrac{7}{10}$

3) $\dfrac{11}{18}$

4) $\dfrac{16}{37}$

5) $\dfrac{2j}{21}$

6) $\dfrac{5k}{7}$

7) $\dfrac{16}{m+1}$

8)   2

9) $o-1$

10) $\dfrac{p+3}{2}$

11) $\dfrac{1}{17}$

12) $\dfrac{3}{10}$

13) $\dfrac{1}{9}$

14) $\dfrac{-6}{37}$

15) $\dfrac{-j}{21}$

16) $\dfrac{2k}{7}$

17) $\dfrac{-2}{m+1}$

18)   2

19) $o-1$

20) $\dfrac{p-4}{2}$

21) $\dfrac{q+2}{2}$

22)   2

23) $\dfrac{3}{s+2}$

24) $\dfrac{x-1}{x}$

25) 4

26) $\dfrac{2}{v-3}$

SECTION 9.5

1) $\dfrac{41}{24}$

2) $\dfrac{49}{40}$

3) $\dfrac{11}{18}$

4) $\dfrac{13}{30}$

5) $\dfrac{31s}{36}$

6) $\dfrac{17t}{20}$

7) $\dfrac{u}{6}$

8) $\dfrac{46v}{21}$

9) $\dfrac{x+w}{wx}$

10) $\dfrac{5z+8y}{yz}$

11) $\dfrac{2(7b-3a)}{21ab}$

12) $\dfrac{2d-5c}{10cd}$

13) $\dfrac{25+h}{5h}$

14) $\dfrac{15+2f}{3f}$

15) $\dfrac{15-4g}{5g}$

16) $\dfrac{h-9}{3h}$

17) $\dfrac{7j+37}{10}$

18) $\dfrac{3k+7}{4}$

19) $\dfrac{4m+11}{45}$

20) $\dfrac{-n-6}{6}$

21) $\dfrac{7p+8}{6}$

22) $\dfrac{-18q+17}{70}$

23) $\dfrac{5r+3}{r^2}$

24) $\dfrac{s^2+2}{s^3}$

25) $\dfrac{4-3t}{t^2}$

26) $\dfrac{u^2-2}{u^3}$

27) $\dfrac{v^4+2v^3+v-1}{v^7}$

28) $\dfrac{w^4-2w^3+w+9}{w^6}$

29) $\dfrac{x^3+x^2-3x+2}{x^4}$

30) $\dfrac{-5y^4+2y^3+y+1}{y^8}$

31) $\dfrac{7z+4}{z(z+1)}$

32) $\dfrac{7a+6}{a(a+2)}$

33) $\dfrac{b-4}{b(b-2)}$

34) $\dfrac{3c+2}{c(c-1)}$

35) $\dfrac{7d+4}{5(d+2)}$

36) $\dfrac{5h+6}{3(h+3)}$

37) $\dfrac{f+12}{4(f-4)}$

38) $\dfrac{-g-3}{4(g-1)}$

39) $\dfrac{5h+9}{(h+1)(h+3)}$

40) $\dfrac{7j+8}{(j-1)(j+2)}$

41) $\dfrac{3k+5}{(k+3)(k+1)}$

42) $\dfrac{3m+7}{(m+3)(m+2)}$

43) $\dfrac{3(n+2)}{(n-2)(n+1)}$

44) $\dfrac{2p-17}{(p+4)(p-1)}$

45) $\dfrac{q-1}{(q+3)(q+1)}$

46) $\dfrac{r-1}{(r+3)(r+2)}$

47) $\dfrac{3s-1}{(s-2)(s-1)}$

48) $\dfrac{1}{t-3}$

49) $\dfrac{u+6}{(u-4)(u-3)}$

50) $\dfrac{2(2v-5)}{(v+4)(v-3)}$

51) $\dfrac{2w+3}{(w+4)(w-4)(w+3)}$

52) $\dfrac{2(x-5)}{(x+1)(x-3)(x+3)}$

53) $\dfrac{-5y+4}{(y+3)(y-2)(y-5)}$

54) $\dfrac{z}{(z-4)(z+3)(z+2)}$

SECTION 9.6

1) $\dfrac{8}{5}$

2) $\dfrac{21}{10}$

3) $\dfrac{2m}{l}$

4) $\dfrac{8q}{7p}$

5) $\dfrac{1}{2r}$

6) $\dfrac{3s}{2}$

7) $\dfrac{t-1}{(t+1)(t-2)}$

8) $\dfrac{u+3}{(u+1)(u-2)}$

9) $\dfrac{(v-1)(v-3)}{(v+1)(v-4)(v+2)}$

10) $\dfrac{(w+5)(w+3)}{(w+4)(w+7)}$

11) $\dfrac{2x-1}{2x+1}$

12) $\dfrac{3y+1}{3y-1}$

13) $z(z+1)$

14) $\dfrac{b}{a-2b}$

SECTION 9.7

1) $j = 8$

2) $k = 480$

3) $\dfrac{1}{2} = l$

4) $\dfrac{9}{2} = m$

5) $\dfrac{-33}{4} = n$

6) $\dfrac{133}{44} = p$

7) $\dfrac{-5}{2} = q$ or $\dfrac{5}{2} = q$

8) $\dfrac{-13}{3} = r$ or $\dfrac{13}{3} = r$

9) $s = 15$

10) $t = 12$

11) No solution

12) $v = 2$

13) $w = -4$

14) $x = -5$

15) No solution.

16) $z = \dfrac{-5}{2}$

17) $a = 6$

18) $b = 4$

19) $c = -3$

20) $d = \dfrac{11}{2}$

21) No solution.

22) $g = \dfrac{1}{2}$ or $g = 6$

23) $h = -4$ or $h = 3$

24) $j = \dfrac{-1}{2}$

SECTION 9.8

1) $f = 4$

2) $g = 5$

3) $h = 32$

4) $j = 3$

5) $k = 3$

6) $l = 10$

7) $m = 3$

8) $n = 8$

9) $p = 6$

10) $q = -3$

11) 384 ounces

12) .75 pounds

13) 6 gallons

14) 11 quarts

15) 4.5 feet

16) 456 inches

17) 3 miles

18) 10560 feet

19) 240 minutes

20) .2 hours

21) .8 minutes

22) 1500 seconds

23) 78 feet

24) 11 yards

25) 1760 yards

26) 43200 seconds

27) 0.07 hours

28) 0.15 hours

SECTION 9.9

1) 20

2) 36

3) 6 and 18

4) 6 and 12

5) 60 mile trip was 4 hours
45 mile trip was 3 hours.

6) Freight: 40 miles per hour
Passenger: 65 miles per hour

7) Car: 40 miles per hour
Truck: 35 miles per hour

8) First portion: 3 hours
Second portion: 2 hours

9) 20 liters

10) 250 minutes

11) 12 gallons

12) 32 liters

13) $16,900

14) $275

15) 30

16) 4 and 12

17) 240 miles: 48 miles per hour
200 miles: 40 miles per hour

18) Day 1: 5 hours
Day 2: 6 hours

SECTION 10.1

1) $2i$

2) $4i$

3) $i\sqrt{51}$

4) $i\sqrt{22}$

5) $36i$

6) $108i$

7) $14i$

8) $-46i$

9) 5

10) 4

11) -10

12) -8

13) 26

14) 2

15) 3 − i

16) 41 + 3i

17) 72 + 32i

18) 6 − 2i

SECTION 10.2

1) -1, 3

2) 1, 6

3) -3, -5

4) -7, 3

5) -2, 6

6) -7, 2

7) -3, 1/2

8) 5/2, 7/2

9) $\dfrac{-19 + i\sqrt{95}}{12}, \dfrac{-19 - i\sqrt{95}}{12}$

10) $6 + \sqrt{71}, 6 - \sqrt{71}$

11) 5, -7

12) -4, -1

13) -5, 2

14) -3, 6

15) 4, 8

16) -3, -5

17) 3, 8

18) -1/3, -2

19) -5/2, 2/3

20) -1, 3/5

21) -5/4

22) 7

23) -3, 2/5

24) 0, 2

25) 0, -8

26) 0, 3

27) -5, 5

28) -9, 9

29) -3, 3

30) -5, -3

31) -5, 1/3

32) 2/3

33) -3/4, 4

34) -5, 0

35) 0, 7

36) 1/8, 5/2

37) 7, -7

38) 8, -8

39) -5, 5

40) -2, 3

41) -5

42) -4

43) $\dfrac{3+\sqrt{109}}{2}, \dfrac{3+\sqrt{109}}{2}$

44) $-2+i, -2-i$

SECTION 10.3

1) 1

2) 1

3) $\dfrac{1}{h}$

4) $\dfrac{1}{j^3}$

5) $\dfrac{1}{8}$

6) $\dfrac{1}{w^9}$

7) $\dfrac{3}{x^{10}}$

8) 27

9) y^5

10) $\dfrac{a^4}{z^6}$

11) $k^9 l$

12) 2

13) $\dfrac{3n^7}{4m^2}$

14) $\dfrac{q}{9p^2}$

15) $\dfrac{1}{8r^2 s}$

SECTION 10.4

1) 2.61×10^1

2) 1.26×10^2

3) 1.263×10^3

4) 6.93×10^0

5) 3.42×10^{-1}

6) 3.52×10^{-3}

7) 6×10^{-1}

8) 2.61126×10^{-4}

9) 611

10) 6120000

11) 38300000

12) 749000000

13) .00000572

14) .000006112

15) .00000001263

16) .0004

SECTION 10.5

1) 8   2) 3   3) 12   4) 6

5) $\left(\sqrt[3]{a}\right)^2$ 6) $\left(\sqrt[5]{b}\right)^2$ 7) $\left(\sqrt[31]{d}\right)^6$ 8) $\left(\sqrt{f}\right)^3$

9) $g^{\frac{5}{2}}$ 10) $h^{\frac{6}{11}}$ 11) $j^{\frac{7}{8}}$ 12) $k^{\frac{8}{3}}$

13) 3125   14) 7776   15) 32   16) 256

17) 32   18) 380204032   19) 1280000000   20) 13824

SECTION 10.6

1) b^{12} 2) F^{24} 3) F^{12} 4) F^{42}

5) $16c^6$ 6) $65536h^{44}$ 7) $59049a^{10}$ 8) $25b^{12}$

9) 256   10) 46656   11) 4096   12) 65536

13) $a^{12}b^{36}$ 14) $c^{24}d^{12}$ 15) $g^{24}h^6$ 16) $j^{15}k^{30}$

17) $4096a^{72}b^{72}$ 18) $46656c^{36}d^{12}$ 19) $32768g^{30}h^{15}$ 20) $19683j^{72}k^{36}$

21) $a^{24}b^{72}c^{72}$ 22) $b^6c^{12}d^{36}$ 23) $p^{72}q^{72}r^{18}$ 24) $r^{18}s^{18}t^{30}$

25) $64a^{12}b^{12}c^4$ 26) $784b^{12}c^{24}d^{32}$ 27) $6859p^{18}q^9r^{24}$ 28) $1156r^{12}s^{18}t^{10}$

INDEX